The Natural History Project was led by Davis Arts Center to celebrate the centennial of the City of Davis in 2017. Davis Arts Center crowd-sourced suggestions for 100 items of natural history significance; through this process, an intense sense of place rooted in the comic (leaf piles, cockroaches and discarded Christmas trees) to the ephemeral (fog and wind) was collected. Over the following year, opportunities for drawing, painting and observing the natural environment and art supplies were provided to participants. Davis Arts Center collected over 600 pieces of artwork from 540 people to compile this book. Participants included school children and adults, individuals and groups of people. In particular we would like to thank the teachers that brought this project to their classes, to Tabatha Yang with the UC Davis Bohart Museum of Entomology for the specimens of insects, Stacey Vetter, the lead on the project for Davis Arts Center and the staff including Nicole Hebert, Liz Ashwood, Sadie Brown, Mike Azevedo, and Shelli Porterfield for keeping the project rolling and Kathi Muto-Brownell for the Graphic Design. This book was produced with support from the City of Davis Civic Arts Grants, Yolo Basin Foundation, Putah Creek Council, and the Davis School Arts Foundation.

Requests for permission should be addressed to:
Davis Arts Center, 1919 F Street, P.O. Box 4340, Davis, CA 95617

"Owl" Cover art by Carrie Ceniseroz

www.davisartscenter.org

AF691610

davis **arts** center

We want to acknowledge the land on which we live and celebrate through this book. For thousands of years, this land has been the home of Patwin people. Today, there are three federally recognized Patwin tribes: Cachil DeHe Band of Wintun Indians of the Colusa Indian Community, Kletsel Dehe Band of Wintun Indians, and Yocha Dehe Wintun Nation.

The Patwin people have remained committed to the stewardship of this land over many centuries. It has been cherished and protected, as elders have instructed the young through generations. We are honored and grateful to be here today on their traditional lands.

INTRODUCTION

What is the *je ne sais quoi* of Davis? For me, it is our community, the nature that surrounds us, it is both in the encompassing drama of the High Sierra and in the everyday. It is the arrival of the almond blossoms in spring, the sweet juice of a vine-ripened tomato, the vibrant pink and orange winter sunsets lighting up the sky over soccer practice, and the bright beautiful sunflowers that dot our landscape during the hot summer. Because we live in such a flat place, we celebrate the sweeping vistas of the Coastal Range to the west and the Sierra Nevada to the east.

We can meet up with a friend at the Farmers Market and have a conversation about the tree that fell in a storm, the owl hooting above the fields or the snow geese flying overhead. It is in the many people who have taken the time to go to the Yolo Bypass and experience the Mexican Free-tailed bat's nightly feeding ritual. It is in Putah Creek, the river that flows from the Berryessa Gap providing habitat for salmon to spawn, after making their unlikely journey through the Delta to arrive home in Yolo County.

We celebrate the bees through art, trees through haiku poems on lawn signs, and chickens through bicycle coop tours.

When thinking about the Natural History Guide, we kept returning to the definition of Natural History:

> The scientific study of animals or plants, especially as concerned with observation rather than experiment, and presented in popular form.

We created this book to spend some time observing our natural world, to generate conversation and to encourage people of all ages and artistic experiences to see and represent the natural history of Davis. We wanted it to celebrate the City of Davis' centennial in 2017. We wanted it to be quirky, like us, and of, by, and for the community. Which is why we invited the community to submit artwork, taking care to include everything we received. We are proud to honor art-makers from kindergartners to nonagenarians.

One word that kept coming up when we were compiling the book was quiddity. Quiddity is another term for the essence of an object, literally its "whatness" or "what it is". This book has come to represent the quiddity of Davis Arts Center. We believe that anyone can make art, that art builds community, and that art isn't simply an outcome but a journey and a celebration of experience.

I hope you enjoy this book for its grounding in our sense of place and our unconventional tribute to our landscape, plants, animals, and insects. I hope it helps you notice our natural world, the one just under our noses in our parks and greenbelts, more intensely and in a new way. Perhaps it will inspire you to make art, observe nature, and reflect on your place in our natural history.

Stacie

Stacie Frerichs
Executive Director

CEANOTHUS • CENTURY PLANT • CHINESE PISTACHE • CHINESE TAL

COTTONWOOD • DAFFODIL • DATURA • DISCARDED CHRISTMAS T

ELDERBERRY • FIG • FLOWERING QUINCE • FOXTAIL GRASS • GINK

ACKBERRY • HEUCHERA • IRIS • JASMINE • LEAF PILE • LEMON • LU

MILKWEED • MISTLETOE • MODESTO ASH • MUSTARD • OLIVE TR

RNAMENTAL PEAR • PENSTEMON • PERSIMMON • PINE • POMEGRAN

POPPY • PURPLE NEEDLE GRASS • RED HOT POKER • REDBUD • REDW

ROSEMARY • SALVIA • SOURGRASS • SUNFLOWER • SYCAMORE • TOM

TOYON • TULE • VALLEY OAK • WALNUT • FISH AND REPTILES • FR

ALMON • TURTLE • MAMMALS • CAT • COW • COYOTE • DOG • HOR

TTER • OPOSSUM • RABBIT • RACCOON • RAT • SQUIRREL • INSECTS • B

LACK WIDOW SPIDER • BUTTERFLY • COCKROACH • CRANE FLY • DEE

DRAGONFLY • FLY • MOSQUITO • MOTH • PRAYING MANTIS • FANTA

RDS • BLACKBIRD • CANADA GEESE • CHICKEN • CROW • DOVE • D

EAGLE • EGRET • FINCH • FLICKERS • HAWK • HERON • HUMMINGB

MAGPIE • MOCKINGBIRD • OWL • SCRUB JAY • STARLING • SWALLO

ANAGER • TURKEY • TURKEY VULTURE • LANDSCAPE • CLOUDS • D

REEZE • FLATNESS • FOG • NORTH WIND • PUTAH CREEK • SUN

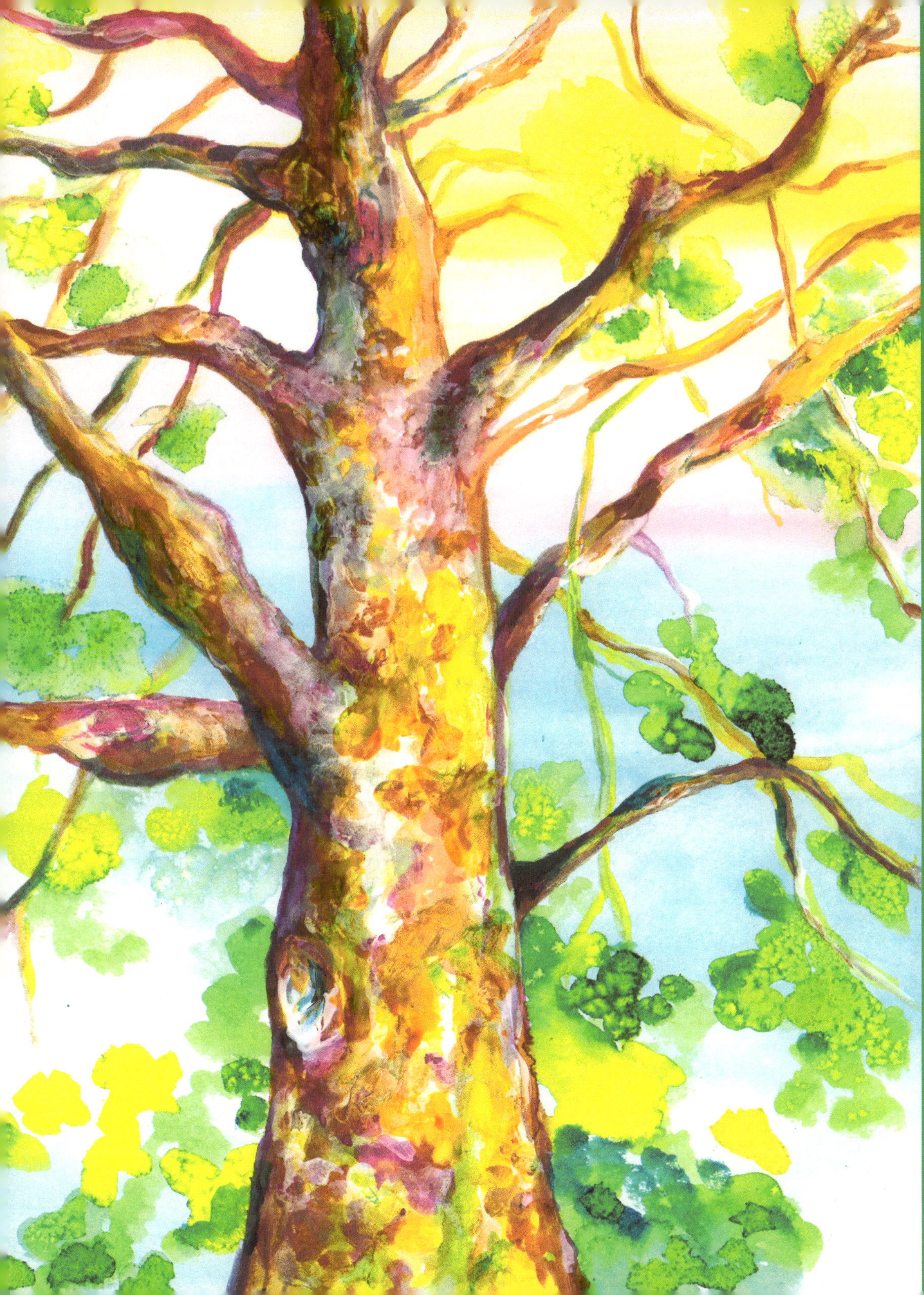

PLANTS

The following artists created the artwork in this section. The list is alphabetical by category and does not necessarily represent the order nor page number the artwork appears; please refer to the page numbers next to the category to locate where the art appears. Some teachers/schools did a class project (CP) for one category; all the artists involved in that project are listed in italics under the teacher and school name within the category. For other class projects the artist chose their category; their names appear under that category with a footnote indicating their teacher/school. If there is a number after an artist's name they provided more than one piece of art for that category.

ACORN (page 7)
Bassel, Claire - CP[1]
Chape, Beth
DeLeon, Christina - CP[1]
Hayes, Julia - CP[1]
McBride, Julie
Parisi, Monica
Perkins, Gabriella - CP[1]
Somah, Suhannah

BERMUDA GRASS (page 10)
Wagner, Jerri

BUCKEYE (page 11)
Williams, Hannah - CP[2]

BULBINE (page 10)
Zamzow, Gary

CEANOTHUS (page 8)
Lincoln, Anne

CENTURY PLANT (page 14-15)
Bekebrede, Heidi
Shepard, Elise - CP[1]

CHINESE PISTACHE (pages 12-13)
Padgett, Susan
CP, Katie Tweltridge/Kelly Valdepeña, June Wood Art Teacher, North Davis Elementary
Abaalkile, Yasir
Carlson, Victorine
Chae, Nathan
Chelius, Lucy
Duan, LingLong
Durbin-Johnson, Olive
Gerhard, Karl
Gomes, Anya
Huang, Licheng
Hubbard, Aden
Lee, Yejin
Martin, Joseph
Mauricio Vasquez Luis, Boris
Nijjar, Armaan
Oerding, Lauren
Reyes, Chloe
Rodden, Finley
Stover, Giuliana
Thomas, Finn
Wang, Jessie
Withrow, Owen

CHINESE TALLOW (page 15)
Ryland, Rebecca

COTTONWOOD (page 15)
Ryland, Rebecca

DAFFODIL (pages 16-17)
Besser, Sally
Candler, Ashlyn - CP[1]
Chiu, Elizabeth - CP[1]
Guerrero, Karla - CP[1]
Rivera, Nicole - CP[1]
Siy, Amanda - CP[1]

DATURA (page 8)
BonDurant, Gayle (2)
Friesen, Julie
Nguyen, Chantal - CP[1]

DISCARDED CHRISTMAS TREE (page 23)
Frerichs. Stacie (3)

ELDERBERRY (page 10)
Casentini, Sophia - CP[1]
Moser, Kiaya - CP[1]

FIG (pages 10-11)
Evans, Ann
Moores, Judy (2)
Wilhelm, Barbara (2)

FLOWERING QUINCE (page 9)
Dailey, Colin
Evans, Ann
Sherwood, Barbara
Young, Sydnie - CP[1]
Zamzow, Gary

FOXTAIL GRASS (page 12)
Berry, Marli - CP[1]
Lindgren, Hannah - CP[1]

GINKGO (page 18)
Fletcher, Ceci - CP[1]
Grehm, Hannah - CP[1]
Kirk, Nancy
Wohlfrom, Judy

HACKBERRY (page 18)
DeLeon, Christina - CP[1]
Privateer, Ann

HEUCHERA (page 12)
Epstein, Peggy

IRIS (page 9)
BonDurant, Gayle
Clark, Bailey - CP[1]
Grehm, Hannah - CP[1]

JASMINE (page 25)
O'Neal, Isabella - CP[1]

LEAF PILE (page 27)
Parisi, Monica

LEMON (page 19)
Evans, Ann
Padgett, Susan
Sherwood, Barbara

LUPINE (page 20)
Maloney, Lori (4)
Wagner, Jerri

MILKWEED (page 15)
Maloney, Lori

MISTLETOE (page 25)
Davis, Samantha - CP[1]
Mudd, Sarah - CP[1]

MODESTO ASH (page 15)
Ryland, Rebecca

MUSTARD (page 33)
Hebert, Nicole

OLIVE TREE (page 24)
Brady, Susan Whitley
Evans, Ann
Maloney, Lori

ORNAMENTAL PEAR (page 21)
CP, Jennifer Aguilar, June Wood Art Teacher, North Davis Elementary
Basharat, Radiatul
Chen, Aiden
Conger, Trinity
Davis, Winter
Durakhanyan, Alek
Foley, Tatiana
Greene, Olijajuan
Gutierrez, Sebastian
Hillman, Paige
Jimenez, Ivonne
Jones, Isabella
Kunkel, Rafael
McCusker, Joseph
Muehlegger, Avery
Pankov, Anton
Petrosillo, Olivia
Raffuse, Truman
Sands, Kina
Singh, Railan
Smith, Jacob
Tao, Kody
Taylor, Akyra
Vogt, Esme

PENSTEMON (page 32)
Carlton, Marsha

PERSIMMON (page 19)
Hunter, Hannah
Sosnick, Galya
Wohlfrom, Judy

PINE (pages 22-23)
Aguero, Nashla - CP[1]
Harrington, Jillian - CP[1]
Jacobsen, Alyssa - CP[1]
Kroener, Cynthia (2)

POMEGRANATE (pages 24-25)
Evans, Ann (2)
Haijan, Soheila
Rice, Emily - CP[1]

POPPY (pages 33-35)
Berry, Malia
Do, Kimberly
Markonic, Vanya - CP[2]
CP, 4th Grade, Teacher: Cammie Portello, June Wood Art Teacher, North Davis Elementary
(Individual artists unknown)

PURPLE NEEDLE GRASS (page 26)
Ryland, Rebecca

RED HOT POKER (page 30)
Fillatti , JoAnne (2)
Seiler, John - CP[2]

REDBUD (page 19)
Vetter, Stacey

REDWOOD (page 23)
Lindroos, Audrey - CP[1]
Post, Sara

ROSEMARY (page 31)
Bedford, Gabby - CP[1]
Maloney, Lori
Morfin, Emma - CP[1]
Wilhelm, Barbara

SALVIA (page 19)
Maloney, Lori

SOURGRASS (page 26)
Ryland, Rebecca

SUNFLOWER (pages 32-33)
Akira, Cherry - CP[1]
Candler, Ashlyn - CP[1]
Holland, Audrey - CP[2]
Houck, Kendall, - CP[1]
Mehta, Jahnavi - CP[1]
Perez, Mari - CP[1]
White, Roni
Ziskind, Linda

SYCAMORE (page 5)
Dale, Yea

TOMATO (page 30)
Candler, Ashlyn - CP[1]
Cota, Ashley - CP[1]
Evans, Ann
Hayes, Lillian - CP[1]

TOYON (page 33)
Harrison, Grace - CP[1]
Ladd, Meghan - CP[1]

TULE (page 26)
Ryland, Rebecca

VALLEY OAK (page 26)
Zamzow, Gary

WALNUT (page 26-29)
Wohlfrom, Judy
Zamzow, Gary
CP, 4th Carrie Ceniseroz, St. James
Addiego, Lucia
Aldrete, Matthew
Alvarez, Miriam
Alvarez, Lena
Antonin, Julia
Bratset, Michael
Bratset, Patrick
Canio, Jayden
Cardwell, Mariana
Carrajal, Maya
Cullison, Jack
Duyre, Jada
Fallon, Thomas
Faust, William
Hansen, Elizabeth
Hedrick, Emma
Hernandez, Sofia
Kim, Charlie
Liu, Ellie
Lopez, Isaac
Mariani, Miles
Miller, Ryan
Nerro, Emma
Palchak, Christopher
Palchak, Katherine
Rodriguez, Sebastian
Rosenbach, Isabella
Simi, Brody
Sousa, Sammy
Stoops, Mason
Tovar, Neta
Yoon, Christopher

[1]Student's Choice Class Project, Teacher: Melanie Bown, St. Francis High School • [2]Student's Choice Class Project, Teacher: Carrie Ceniseroz, St. James School

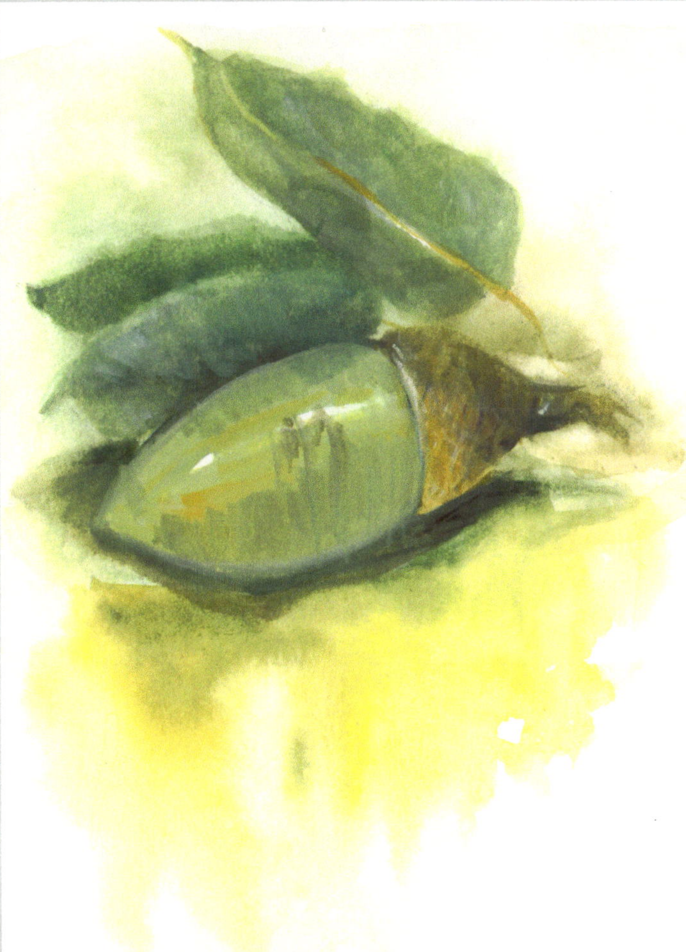

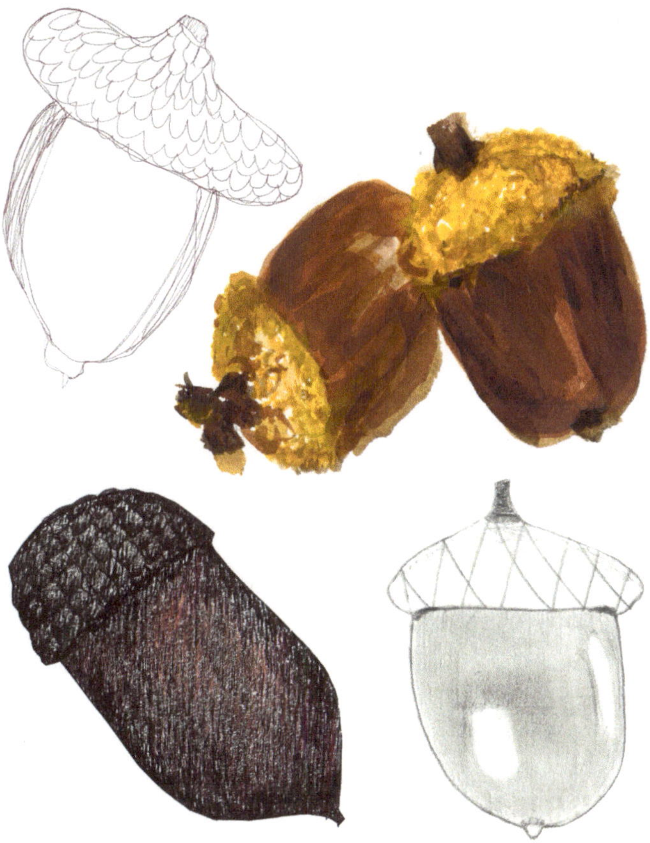

Quercus lobata

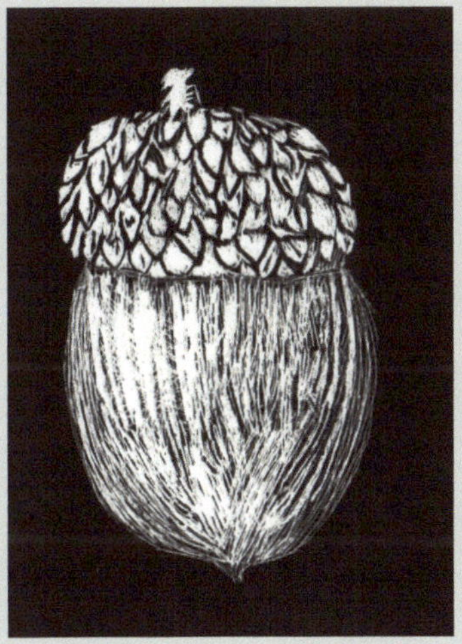

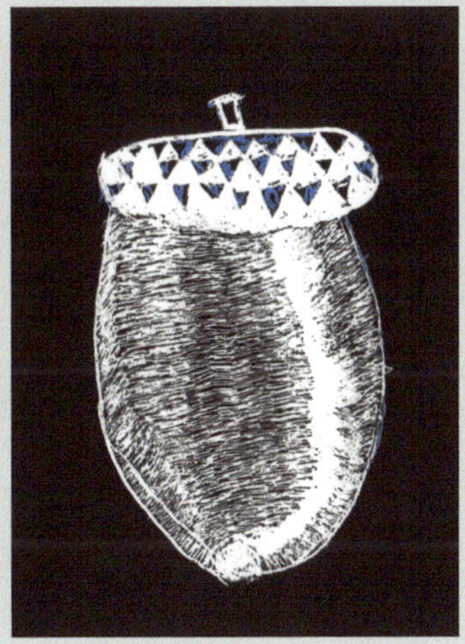

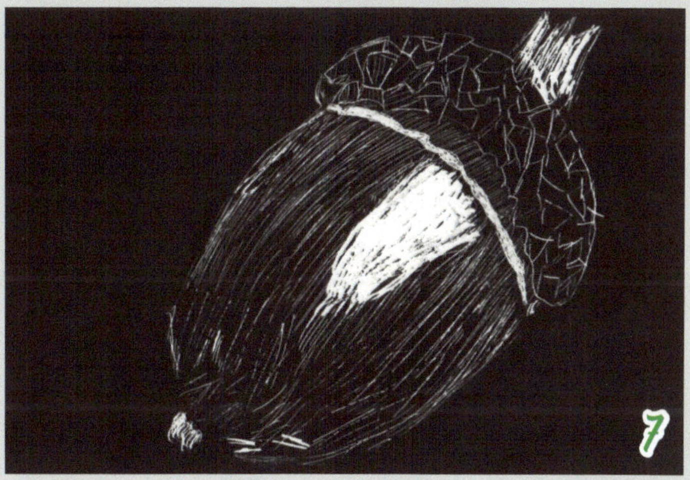

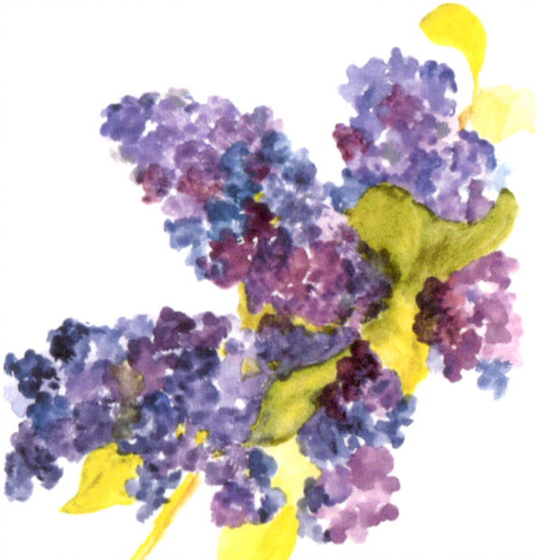
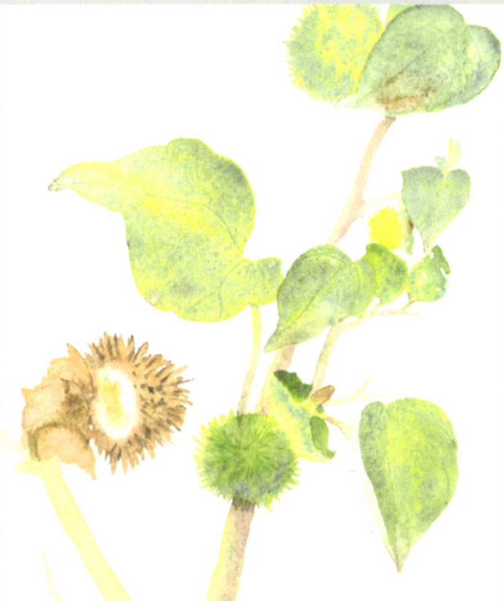
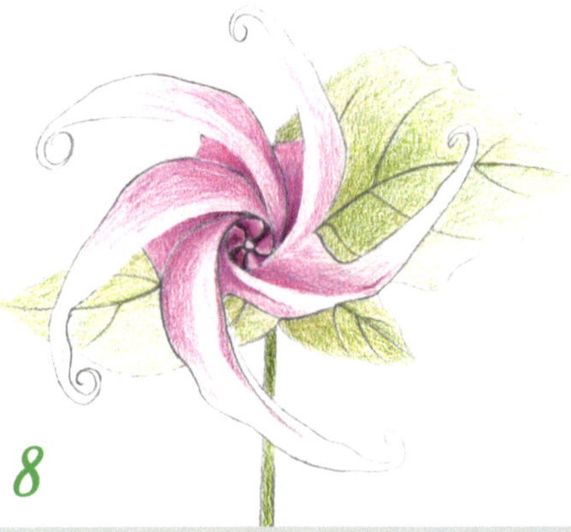
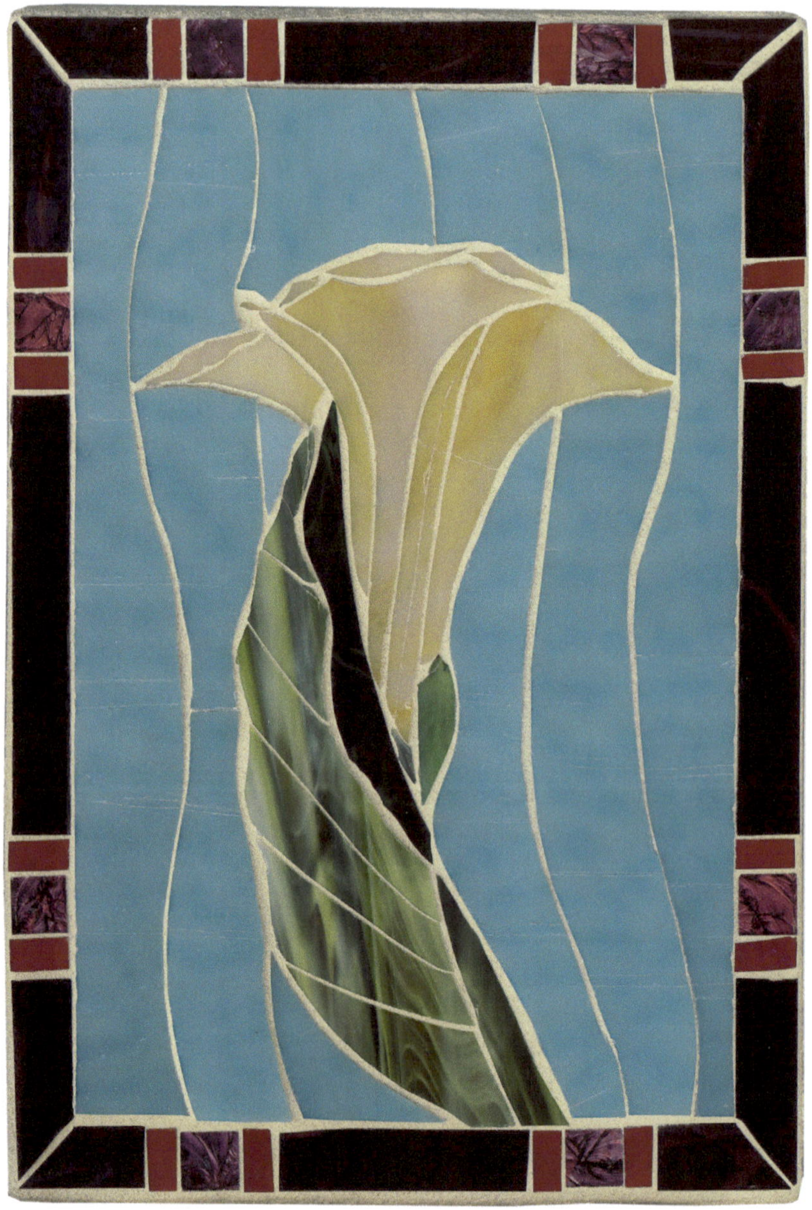

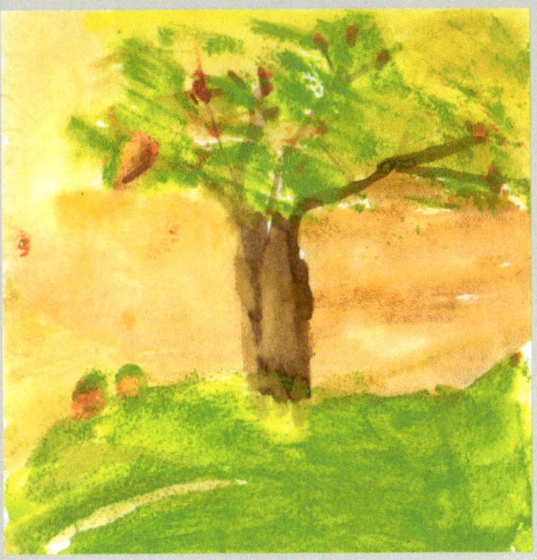
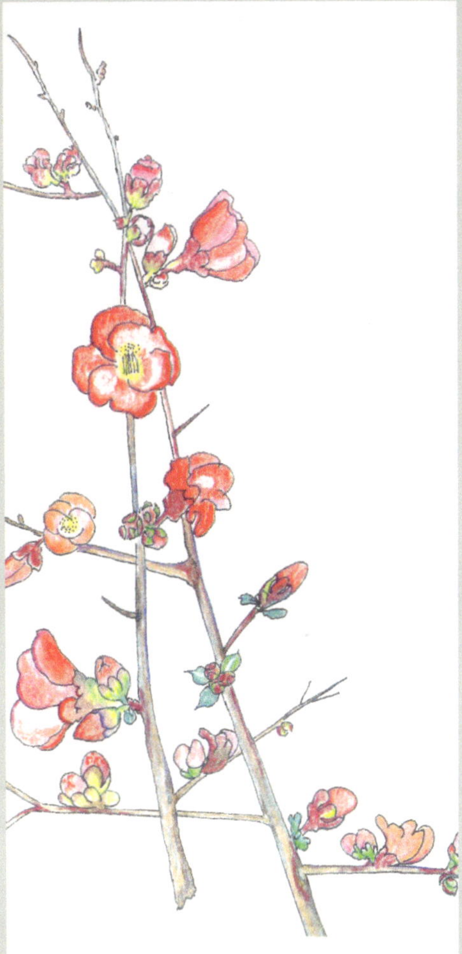
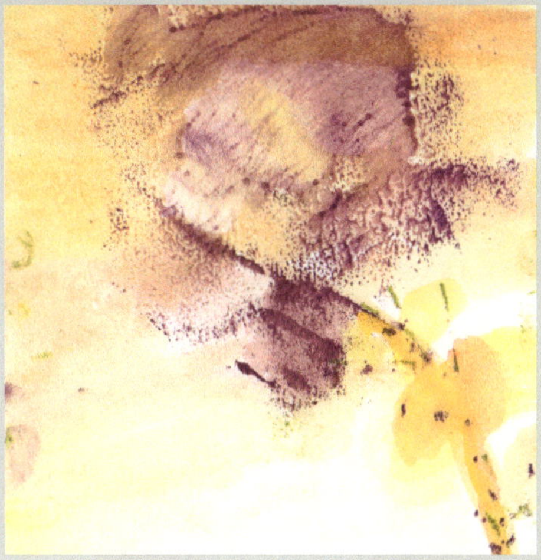
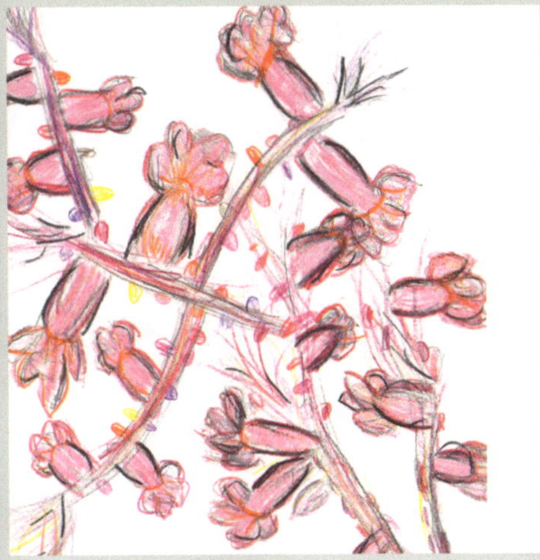
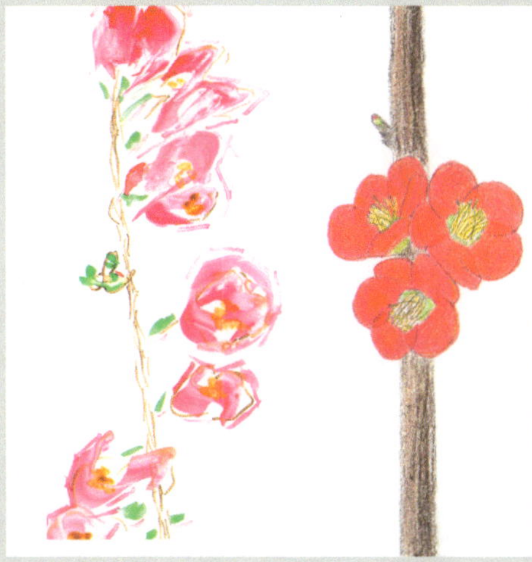
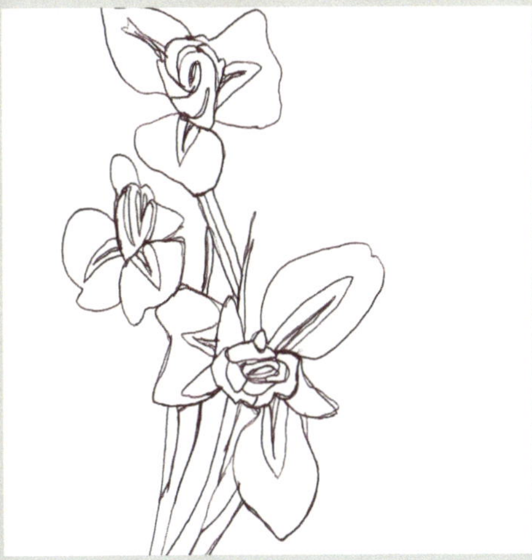
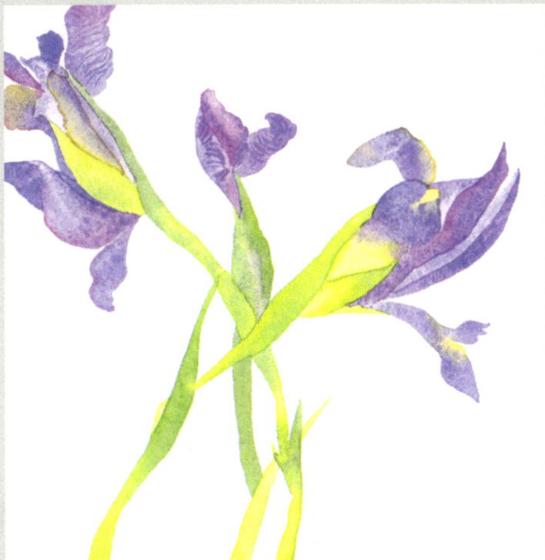
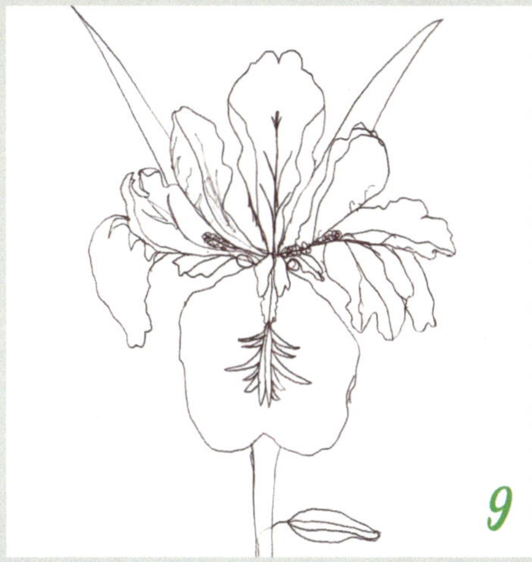

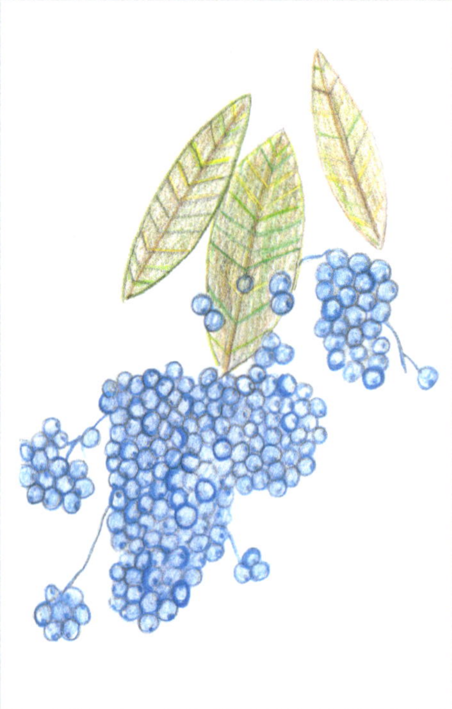

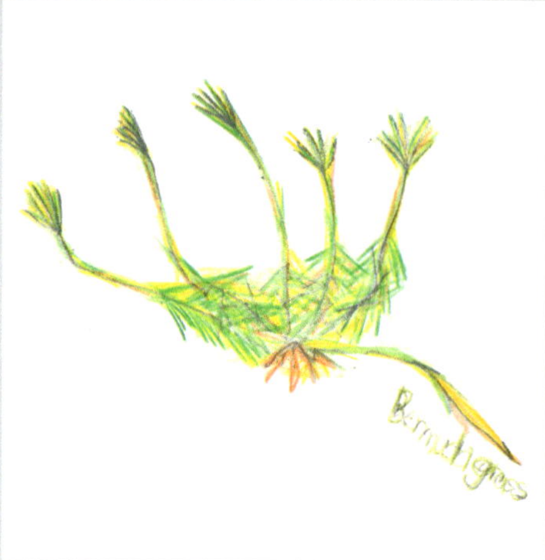

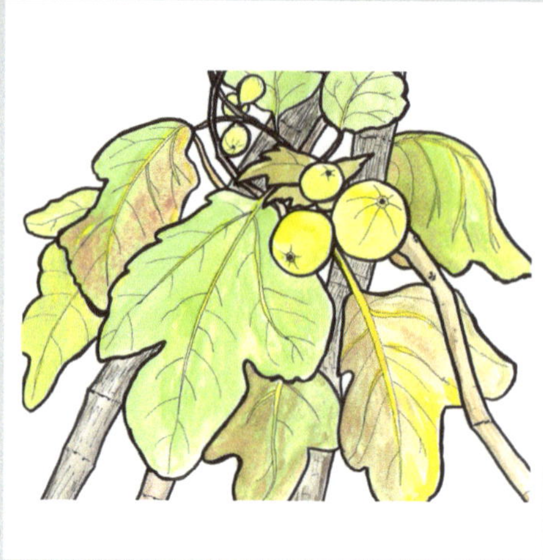

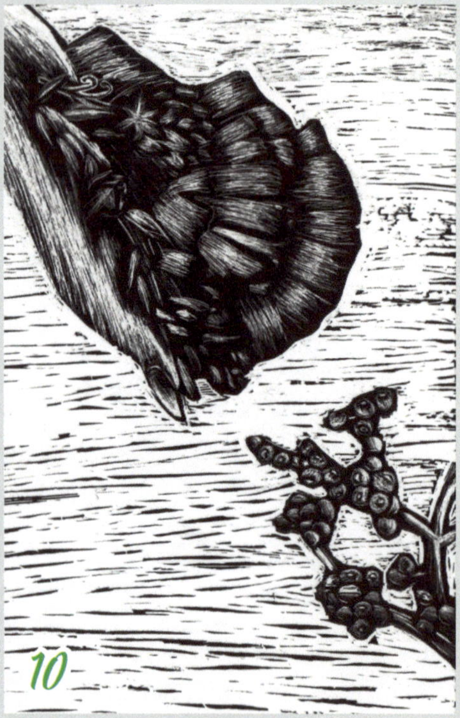

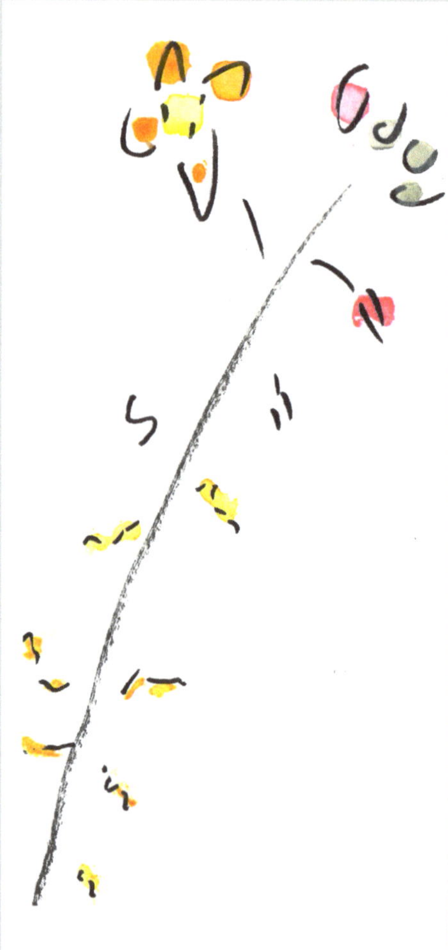

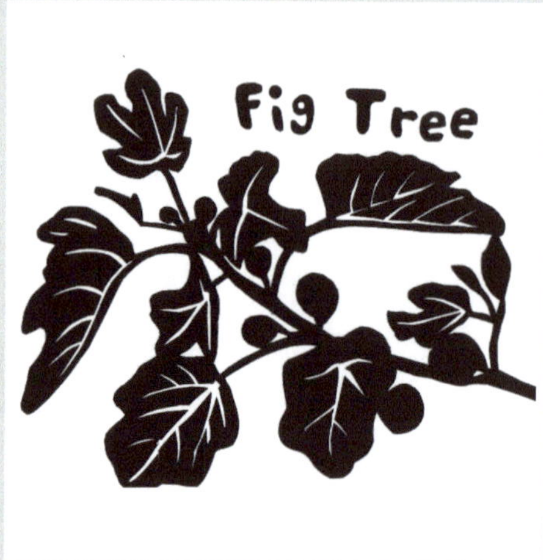
Fig Tree

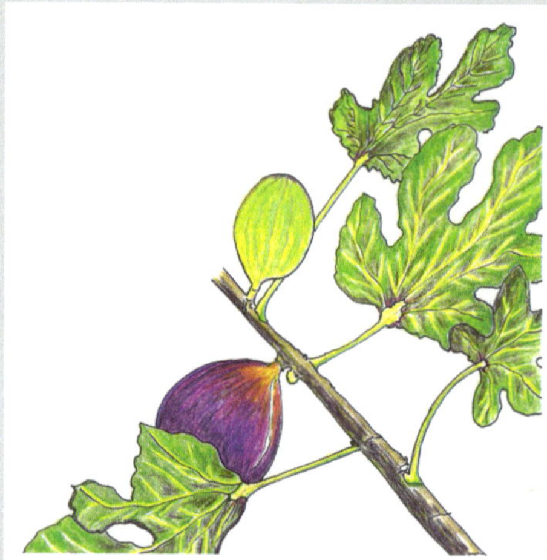

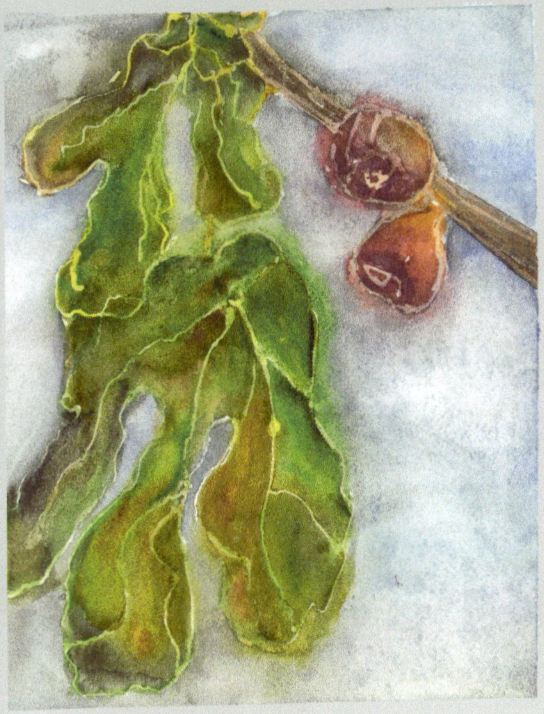

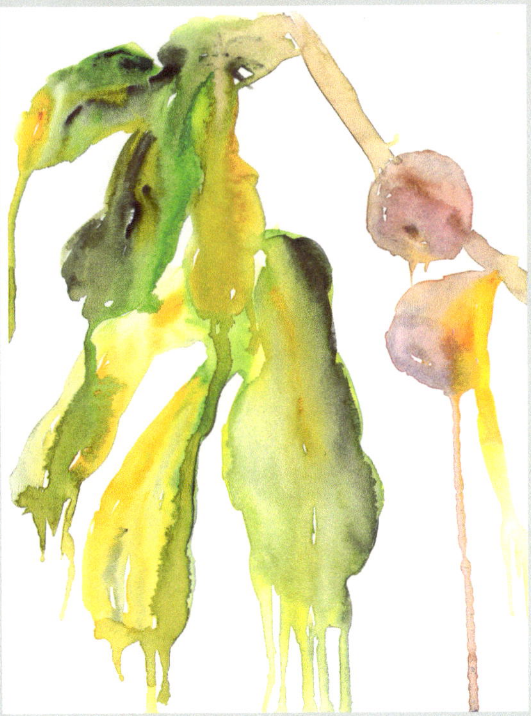

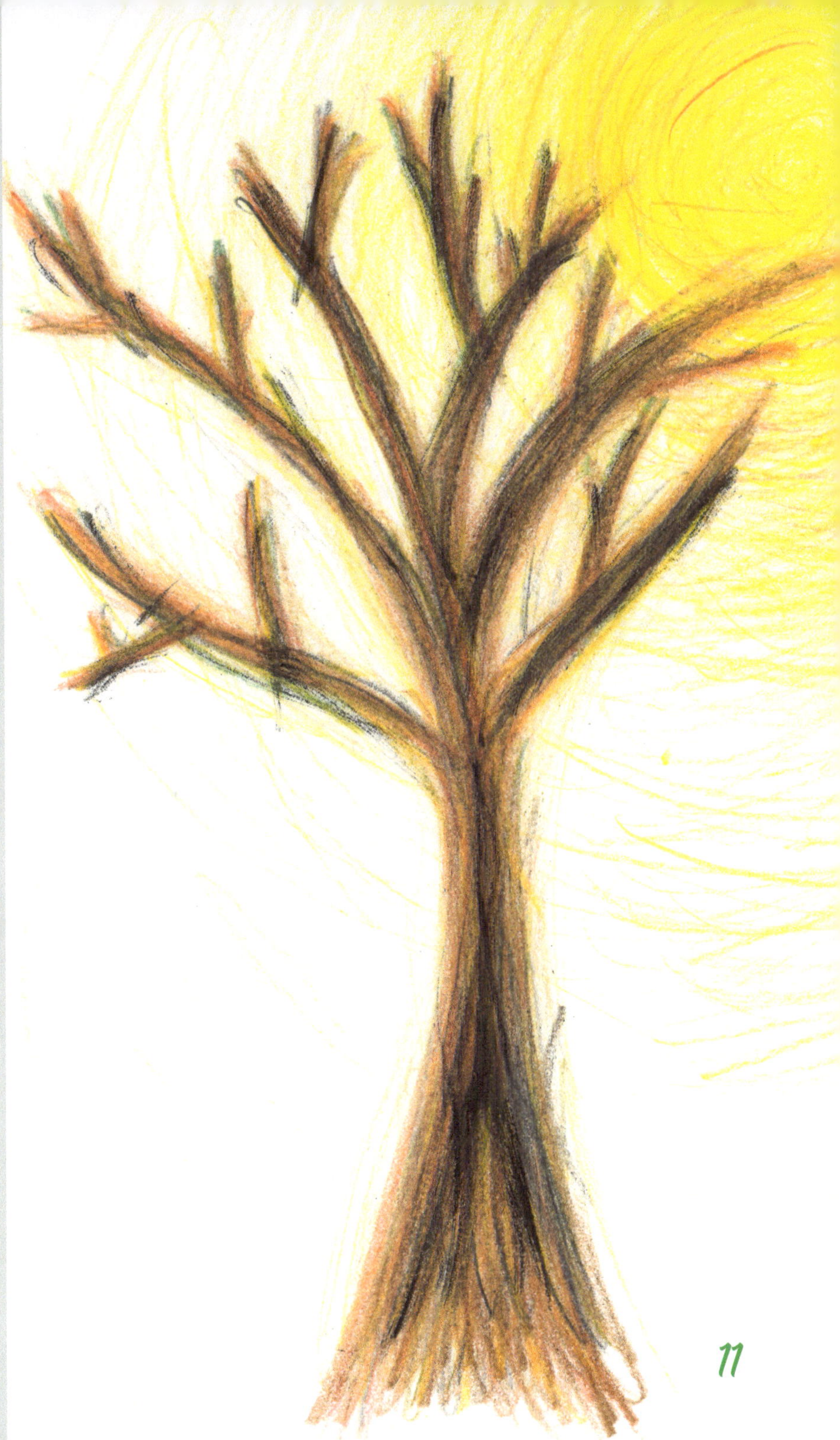

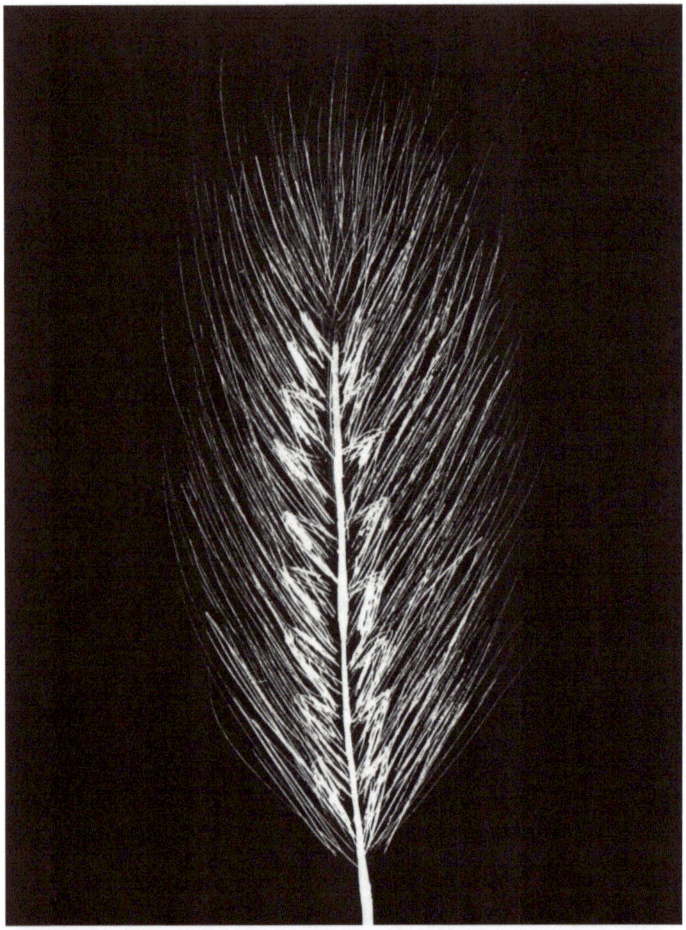

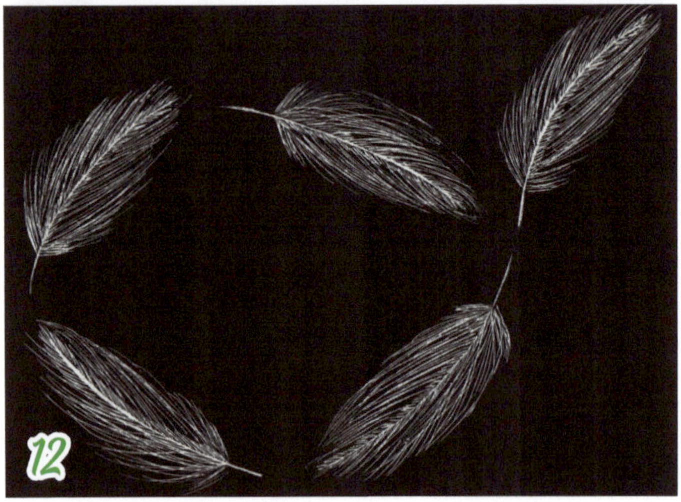

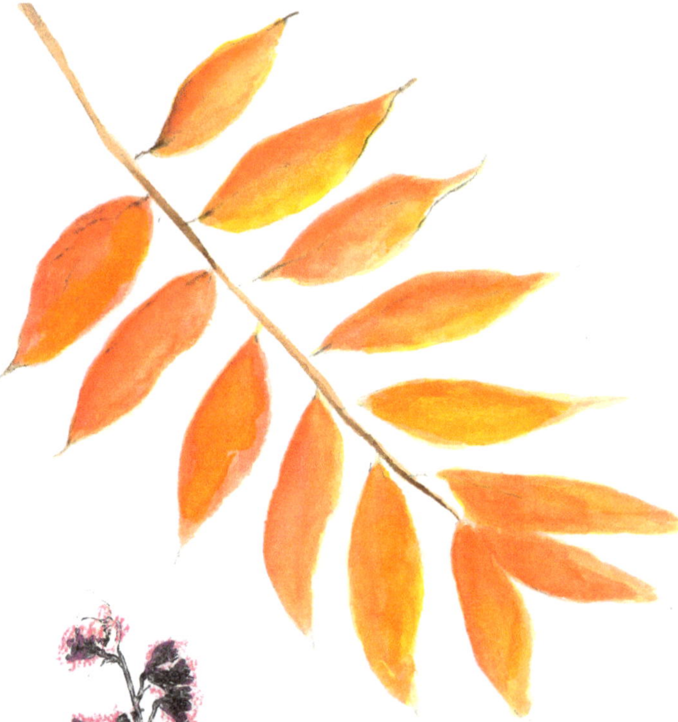

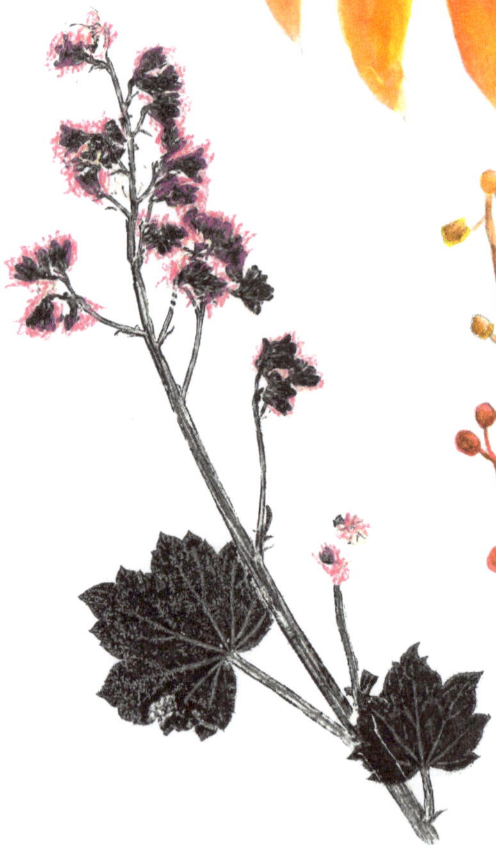

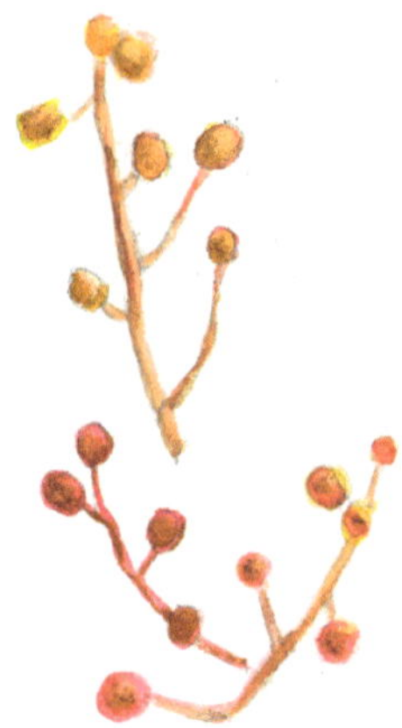

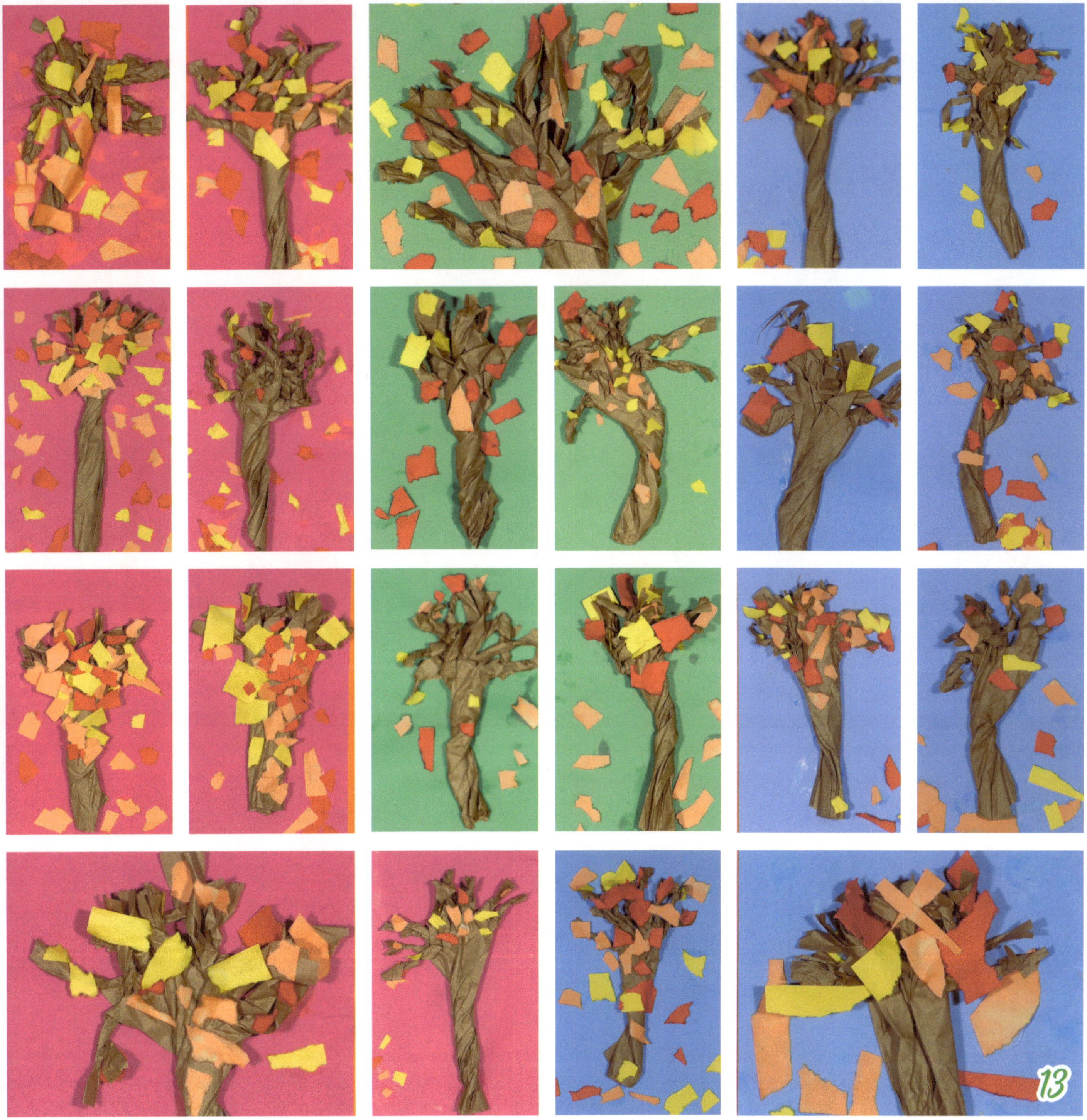

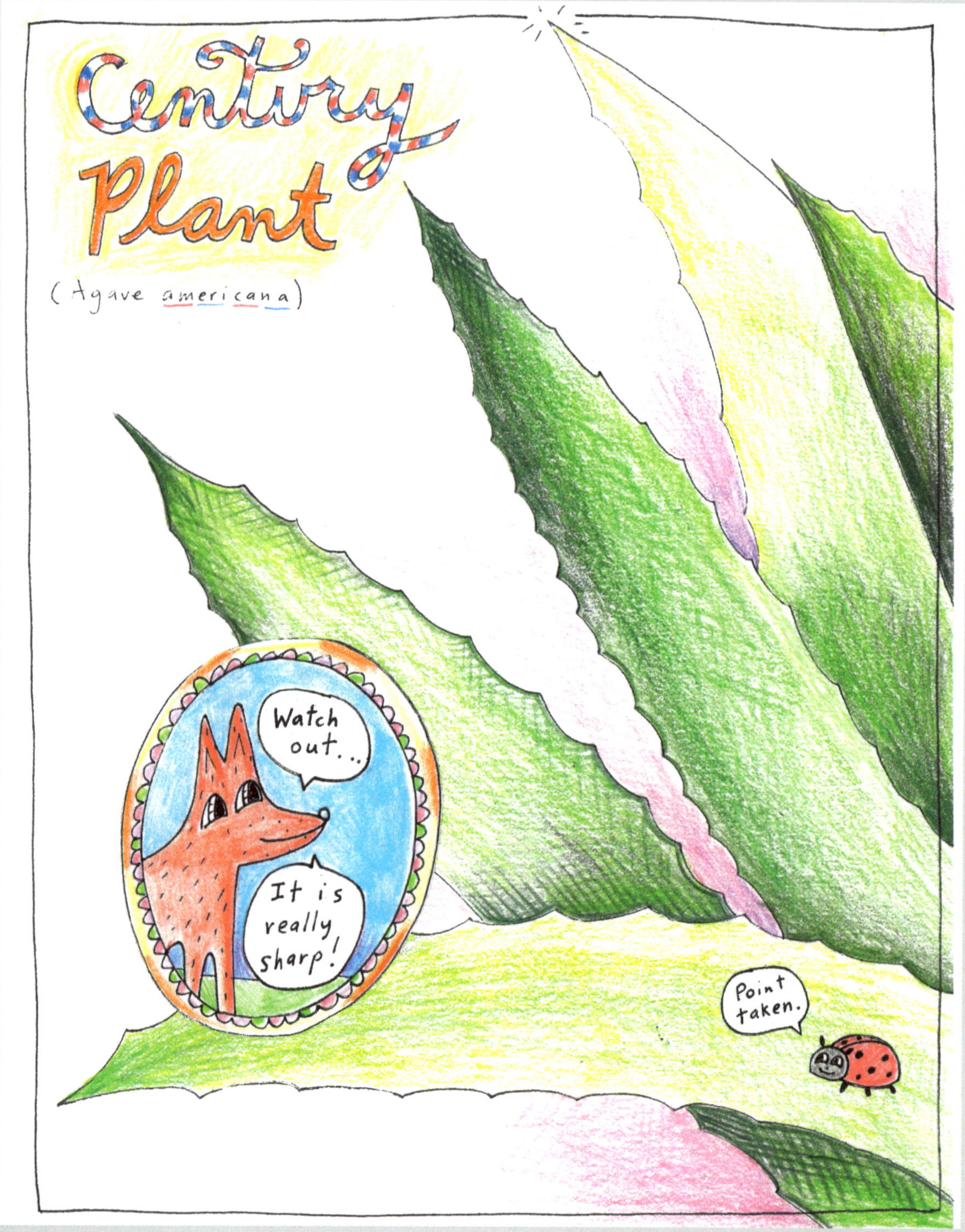
Century Plant
(Agave americana)
Watch out...
It is really sharp!
Point taken.

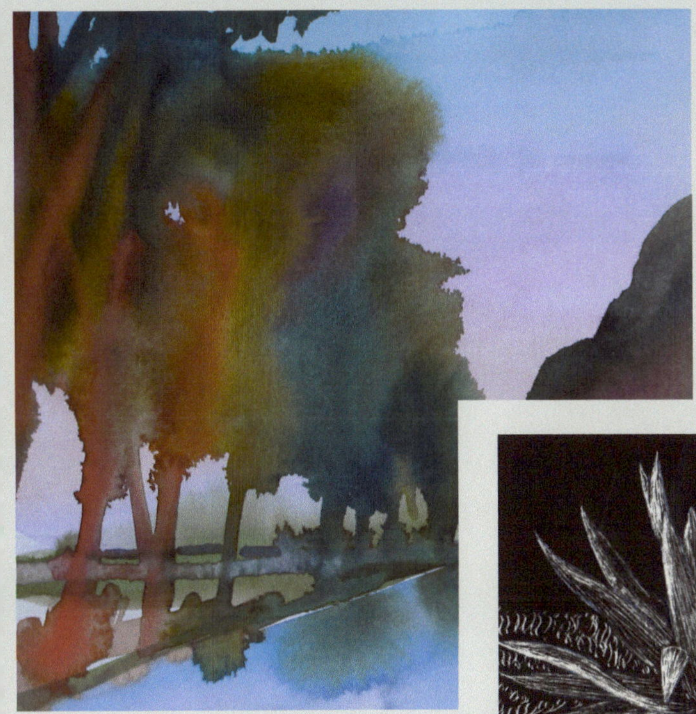

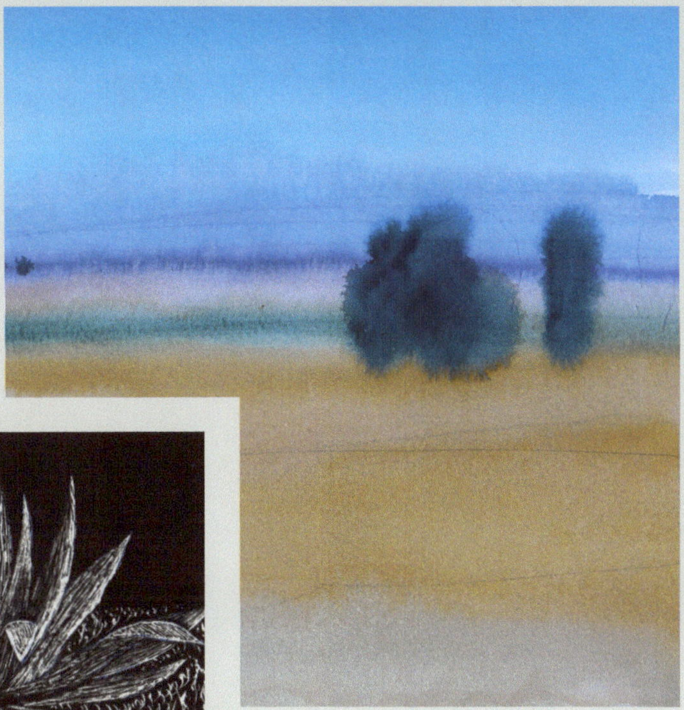

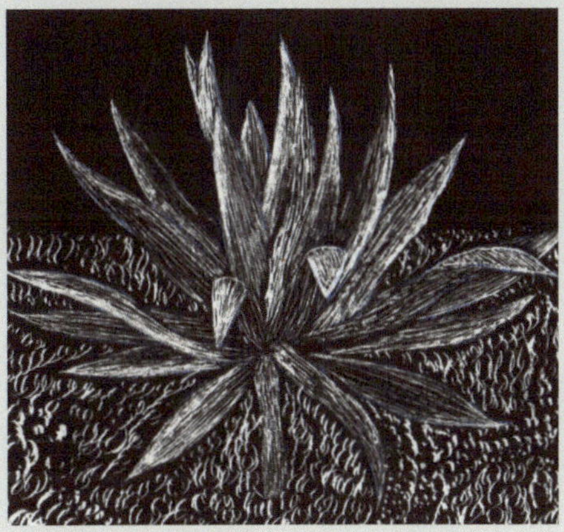

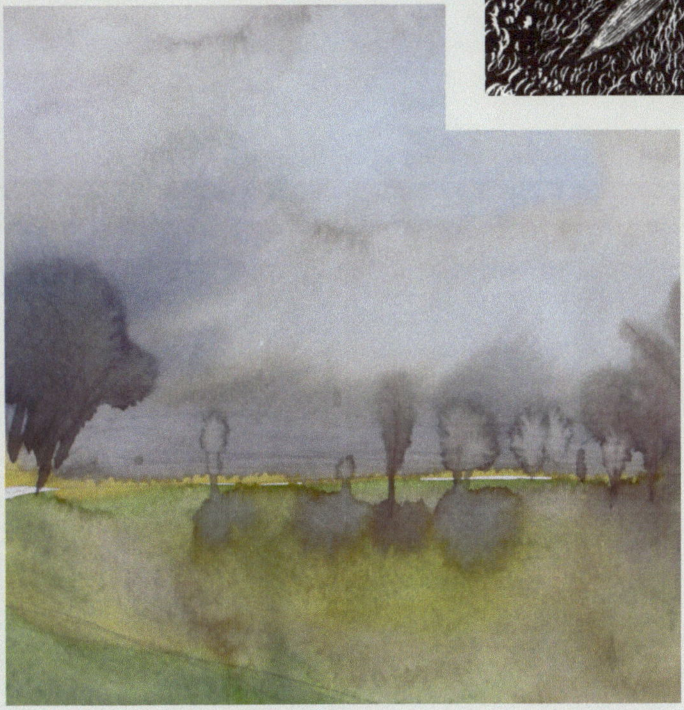

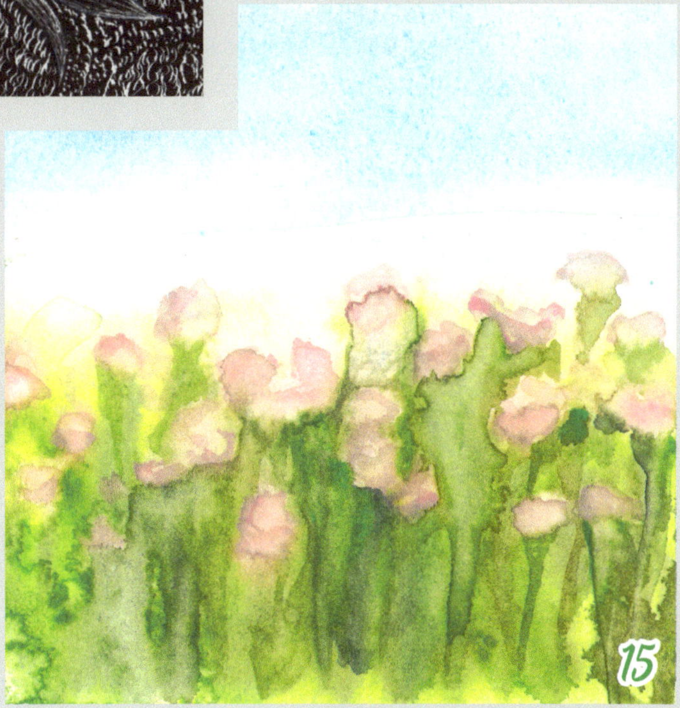

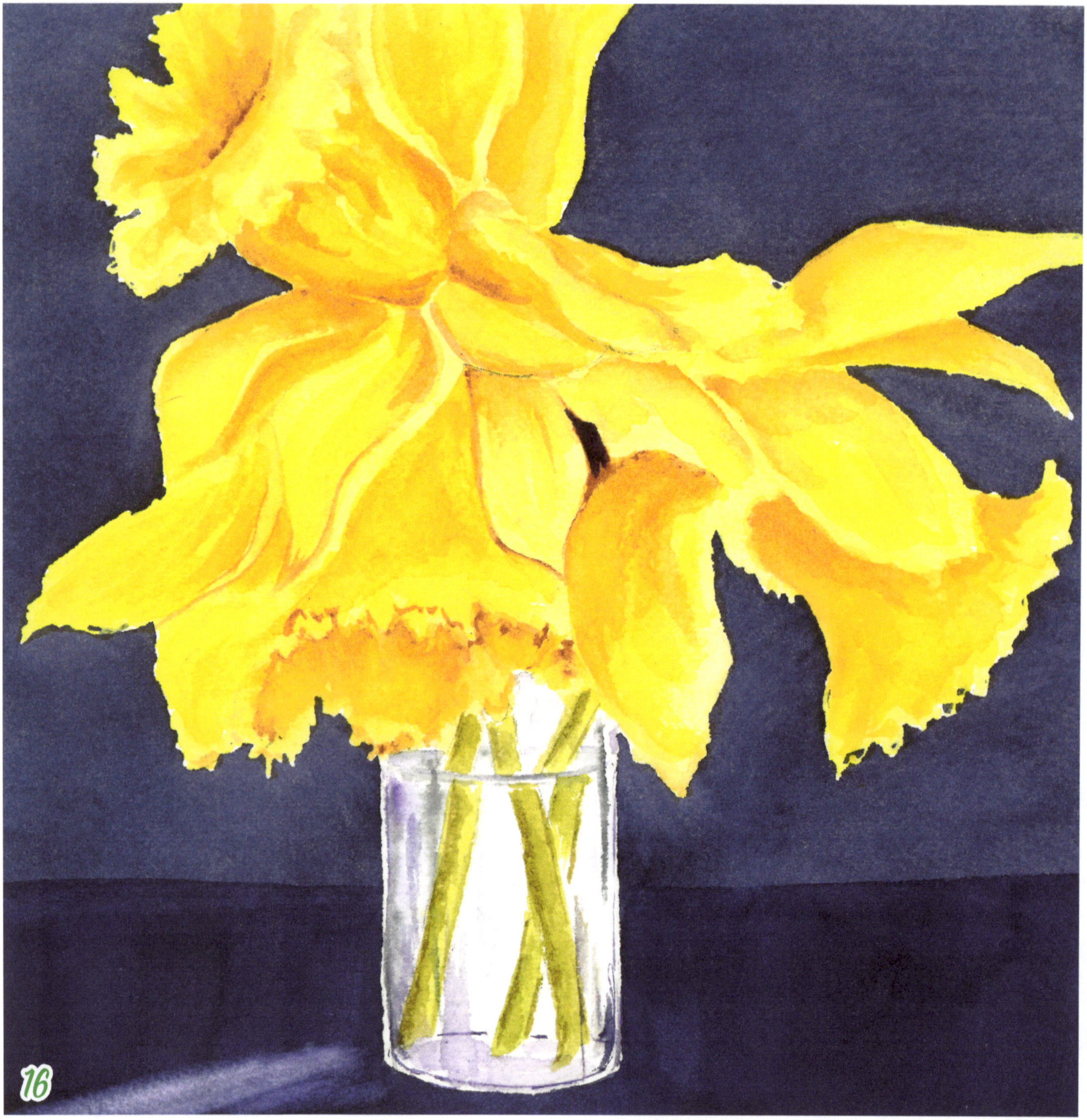

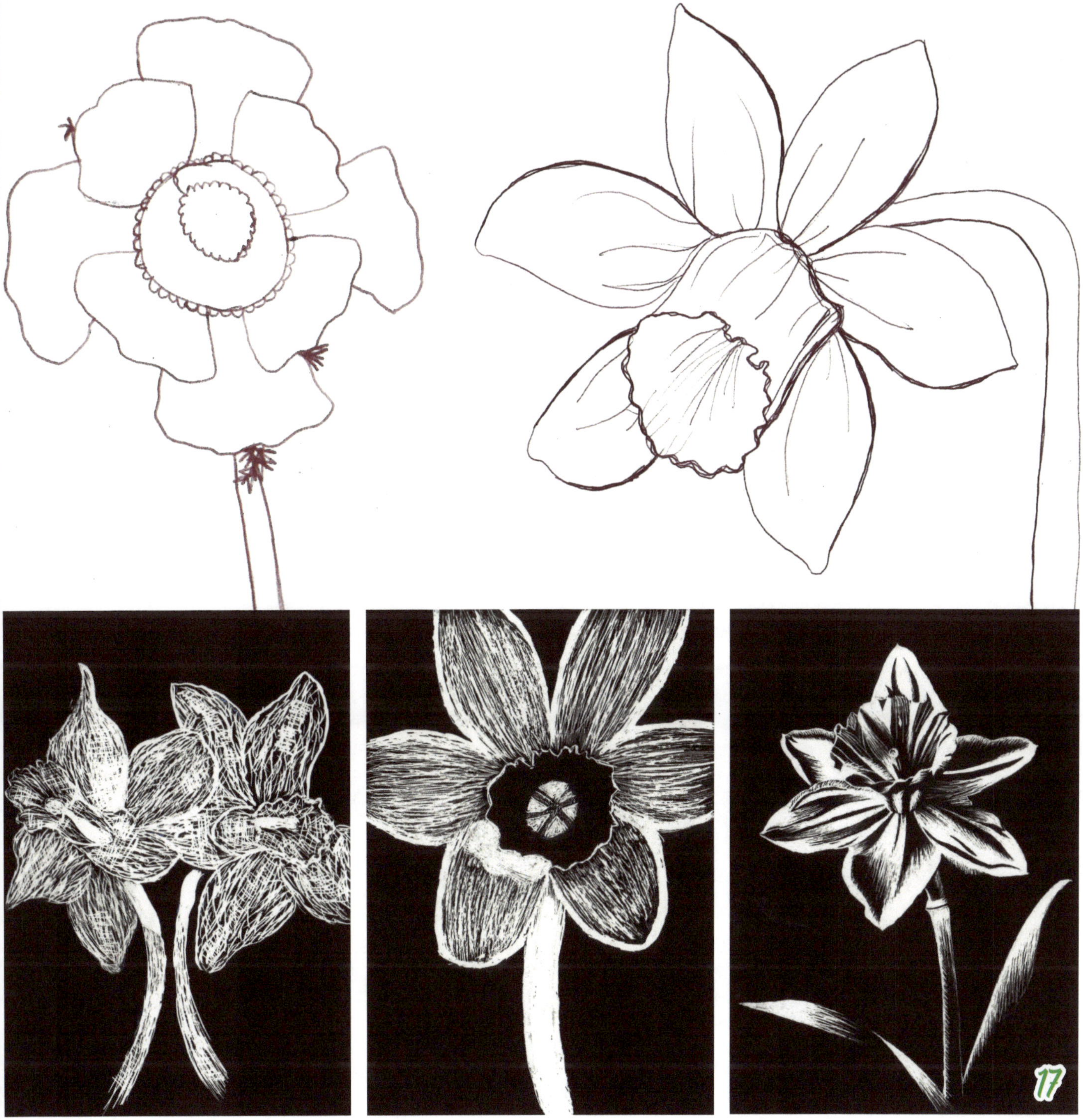

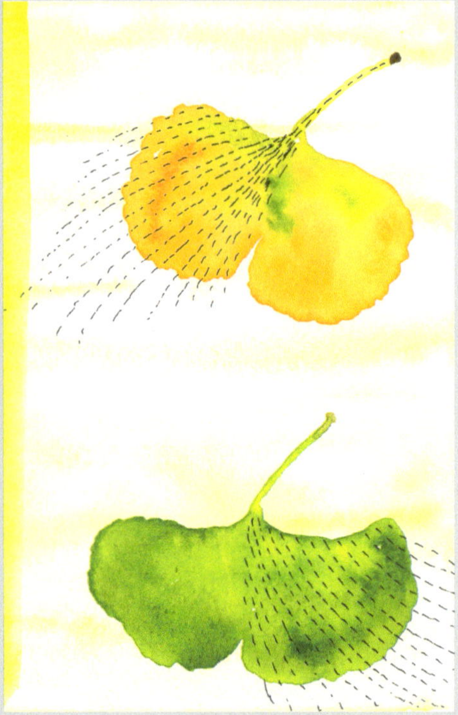
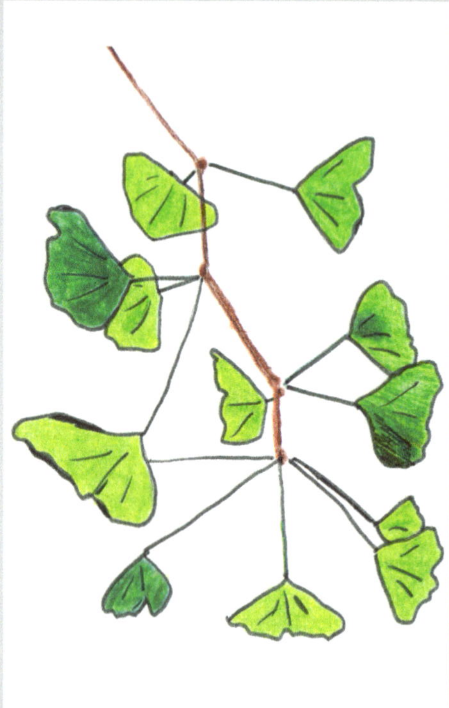
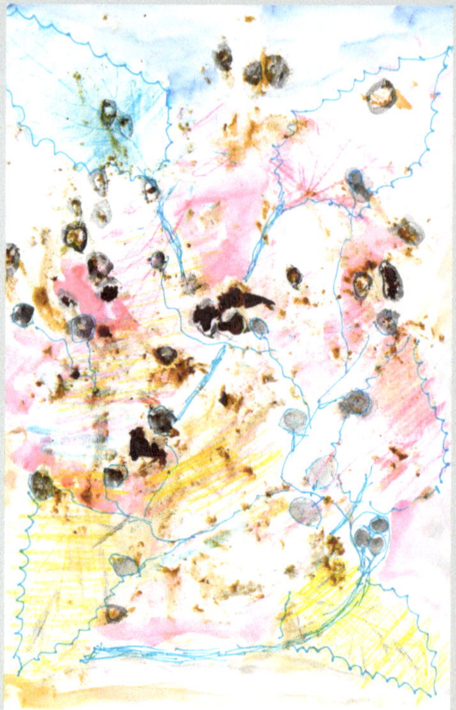
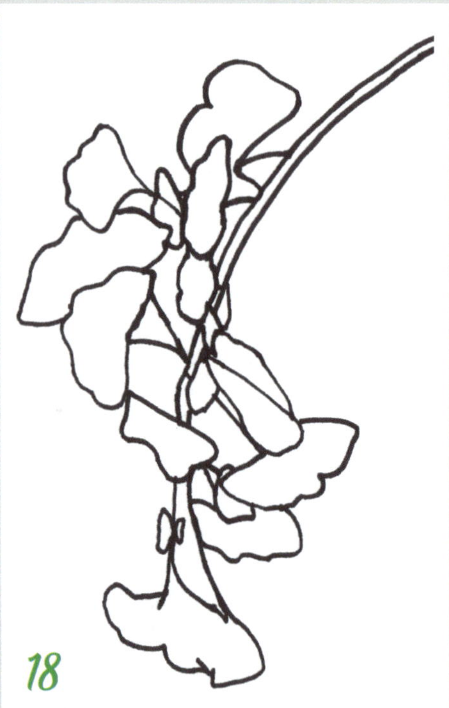
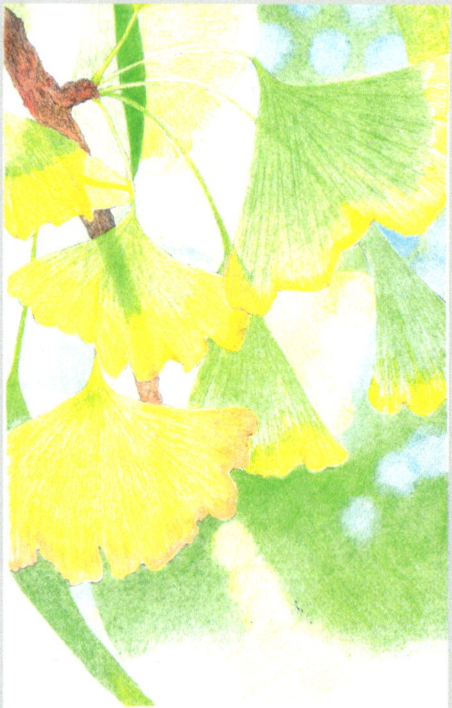
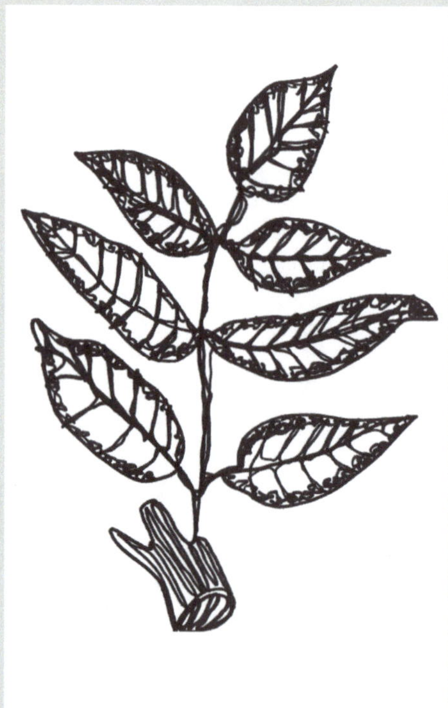

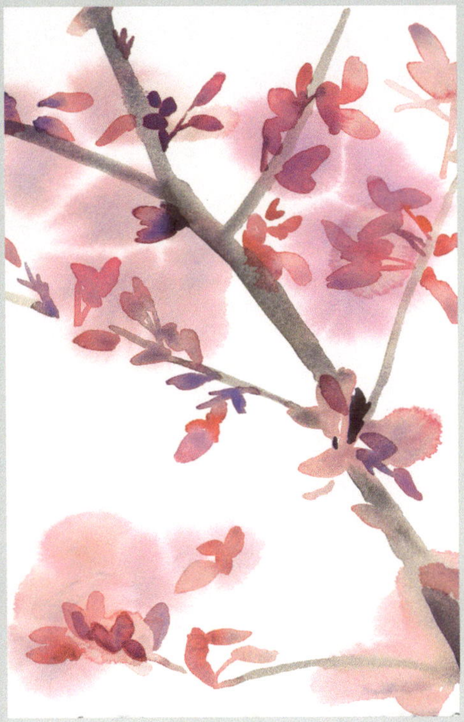

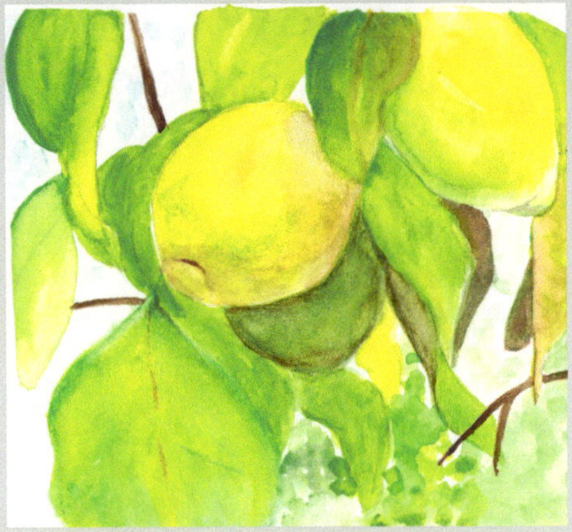

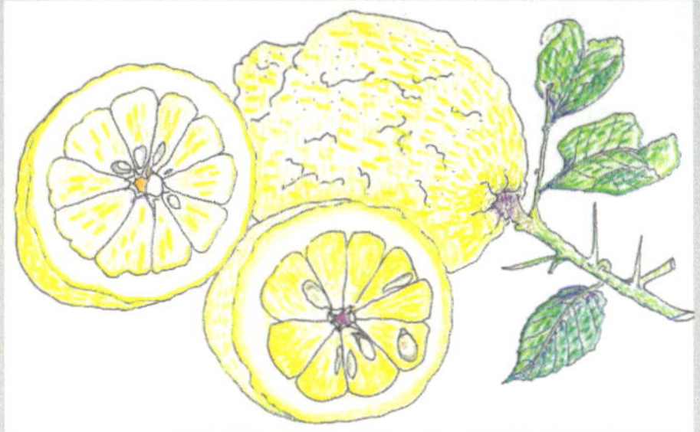

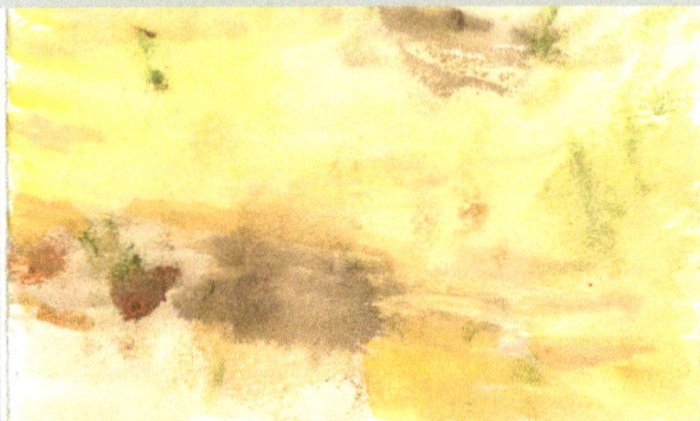

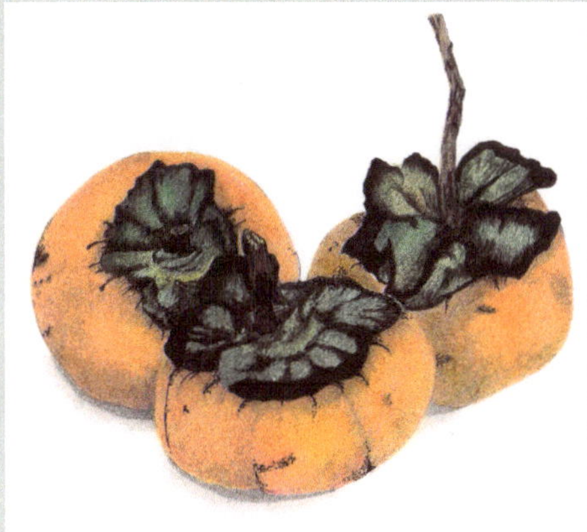

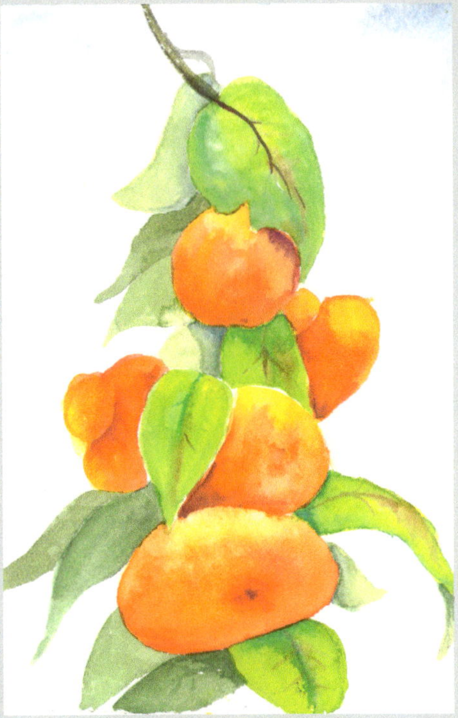

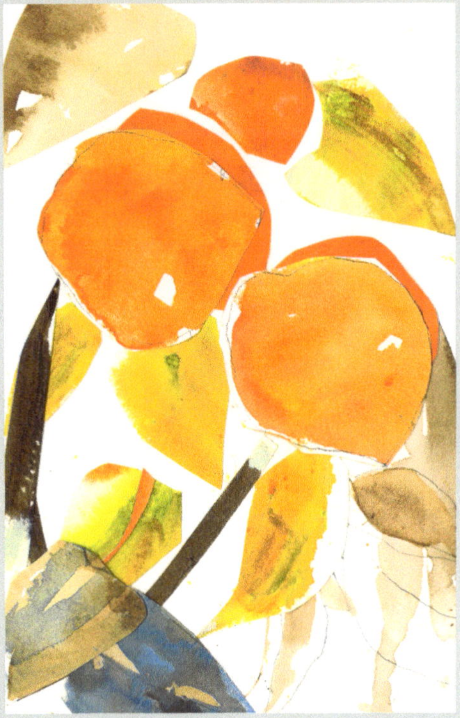

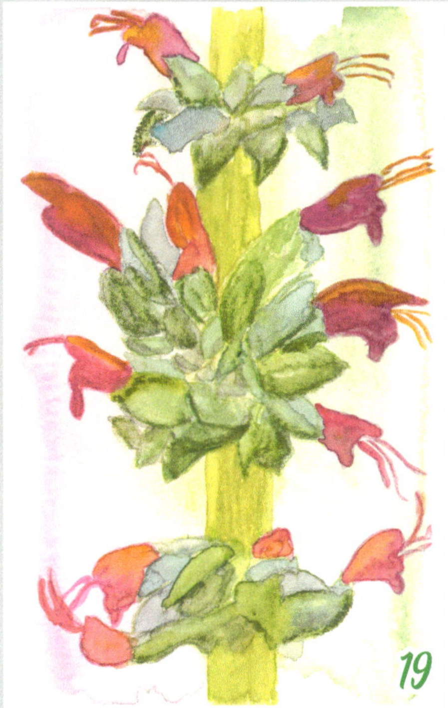

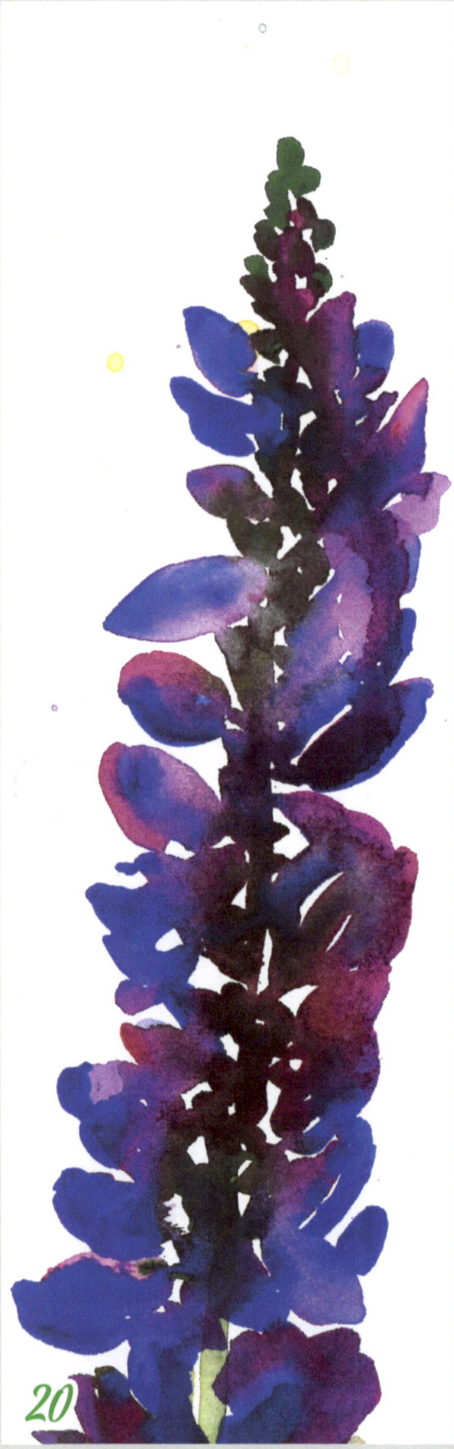

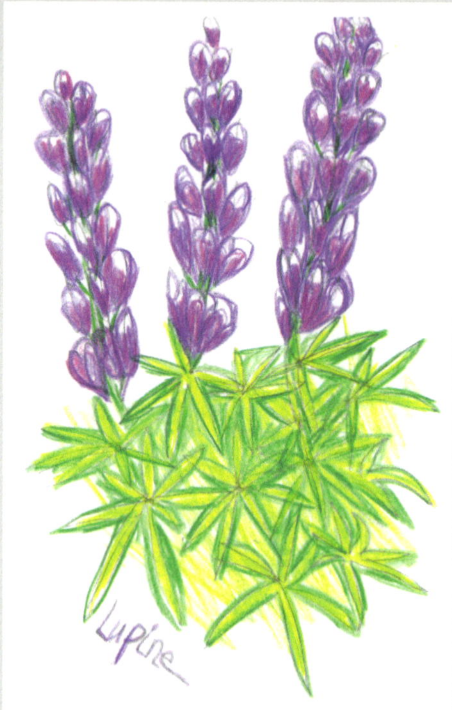
Lupine

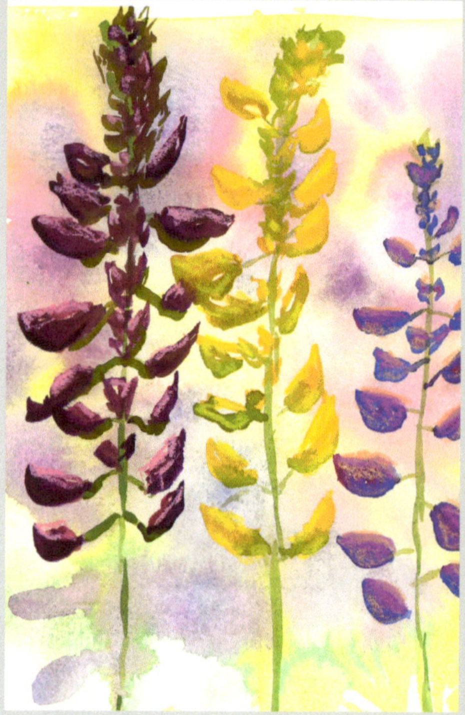

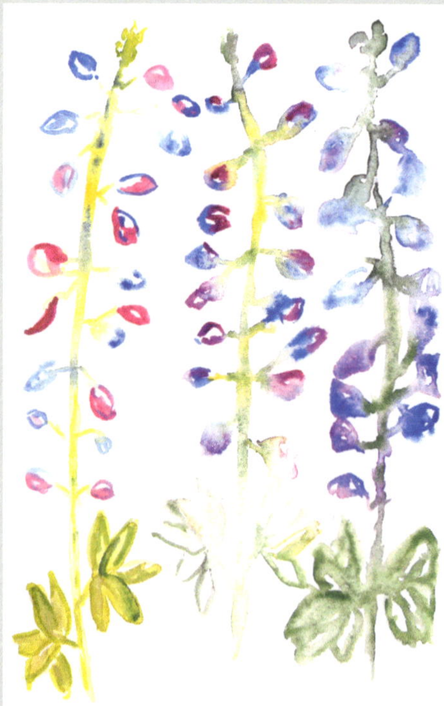

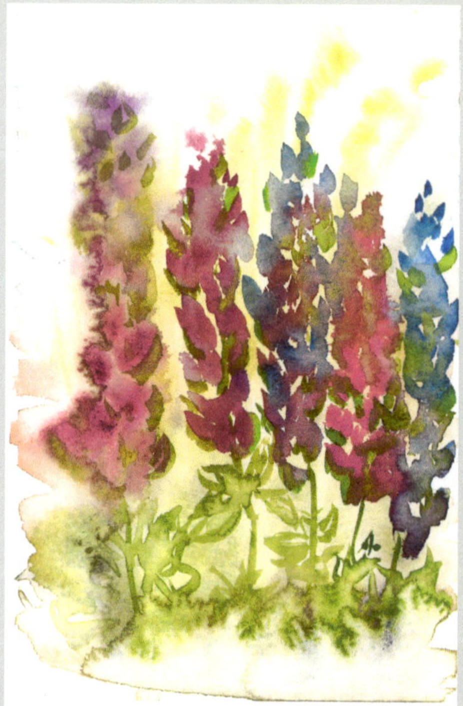

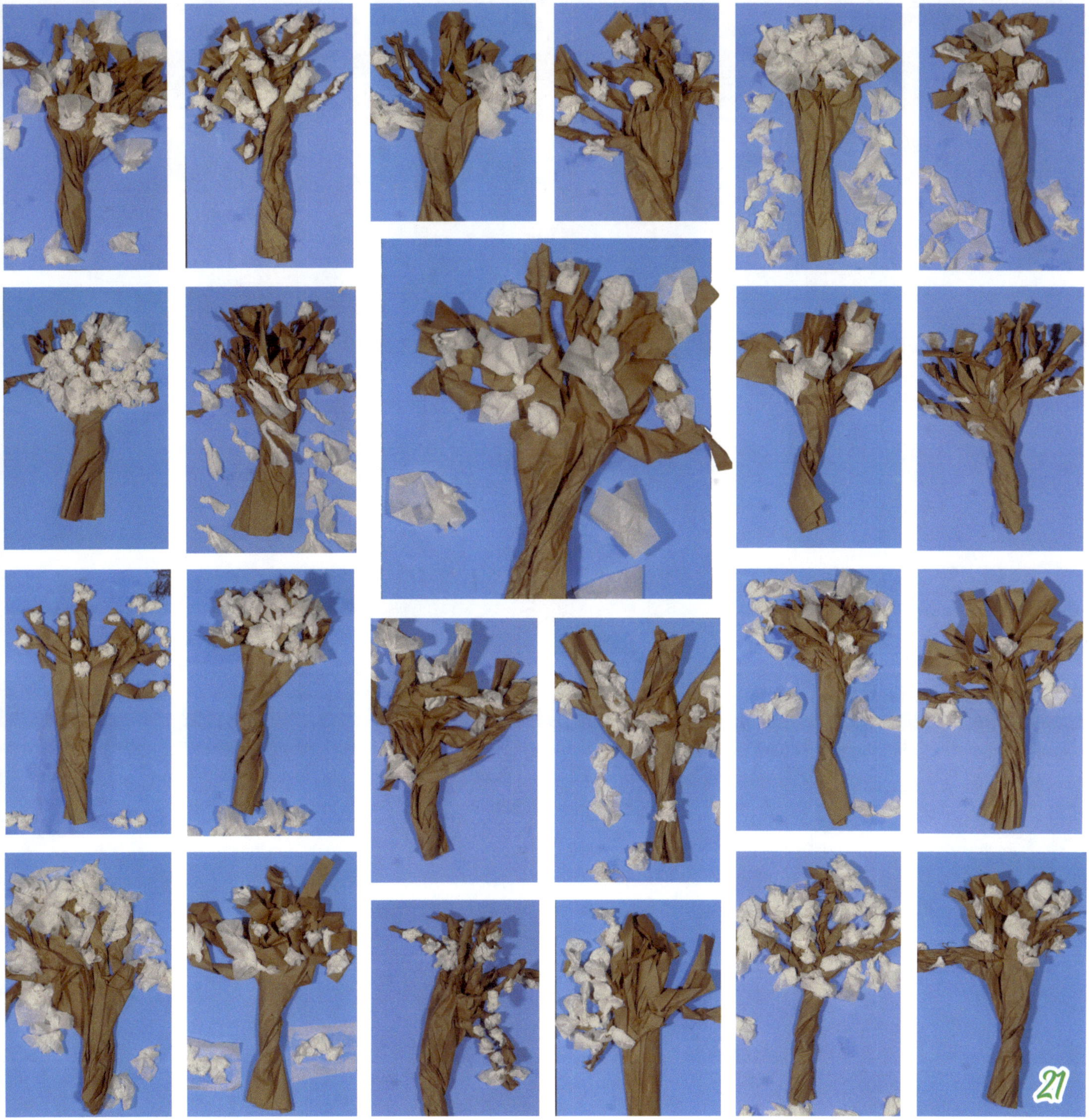

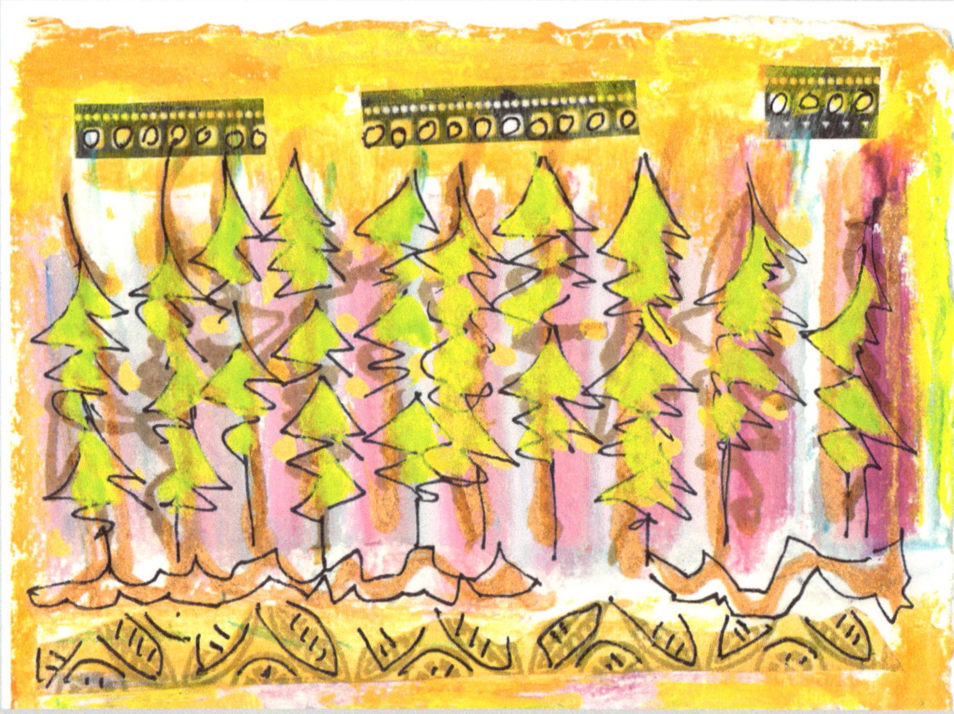

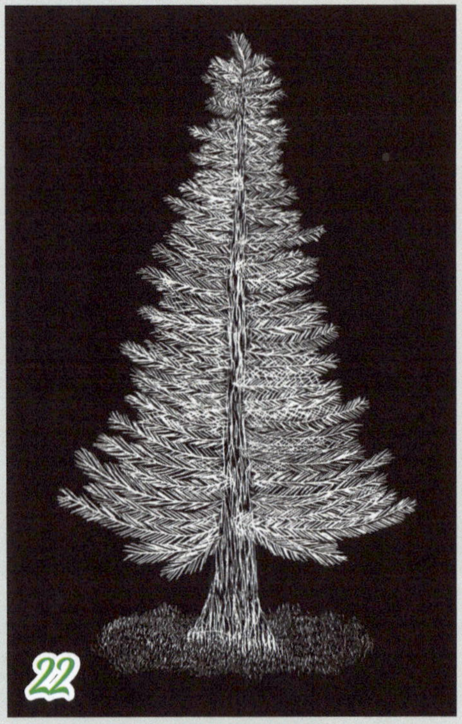

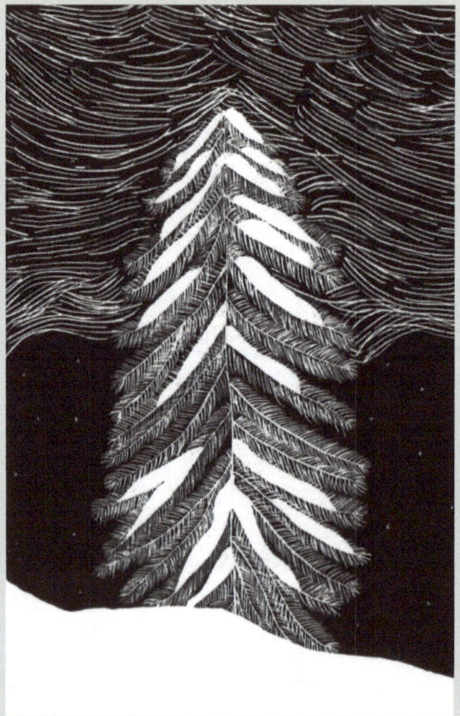

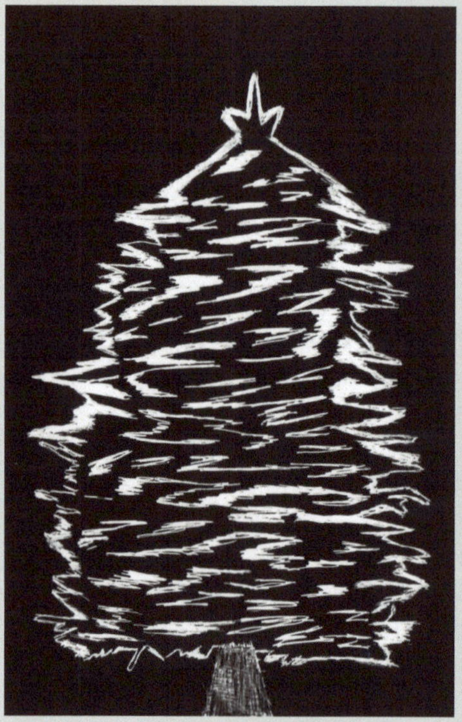

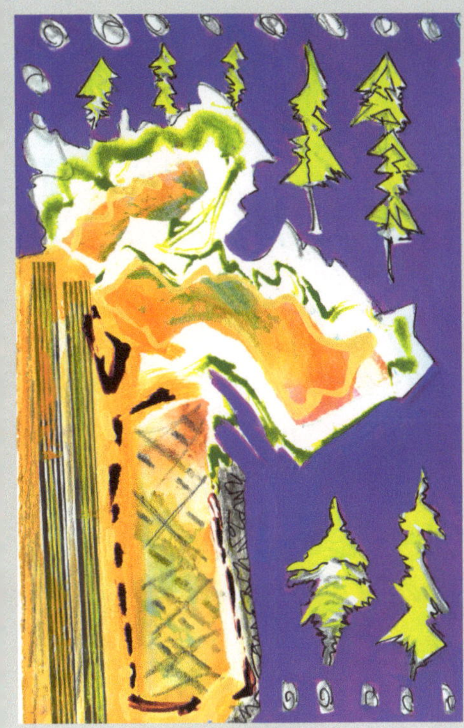
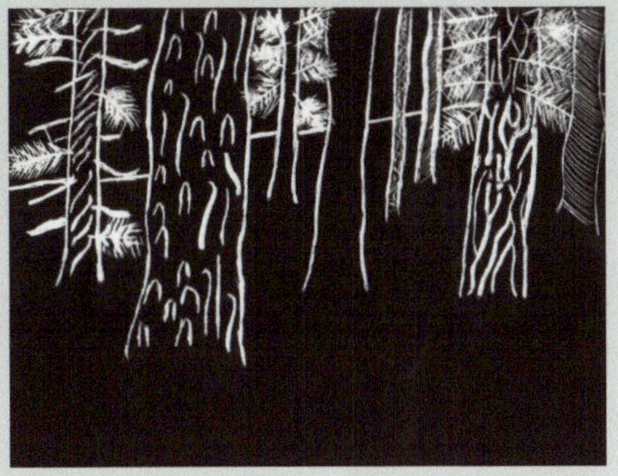
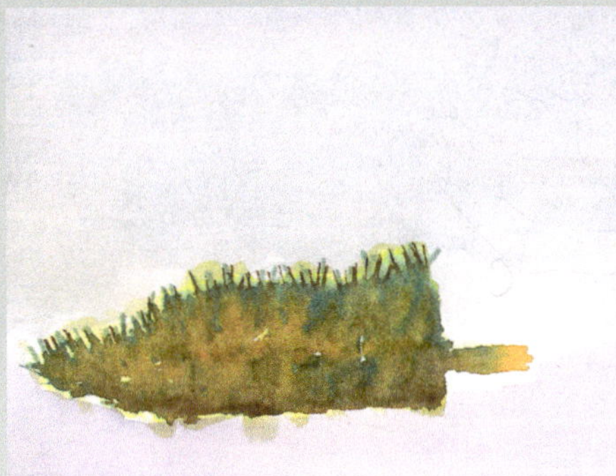
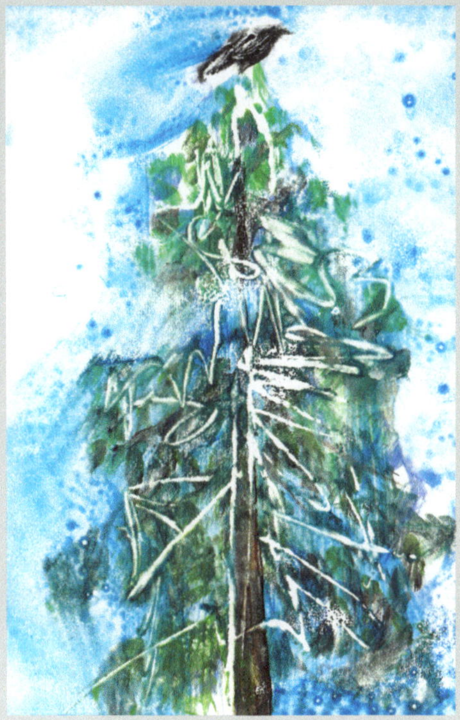
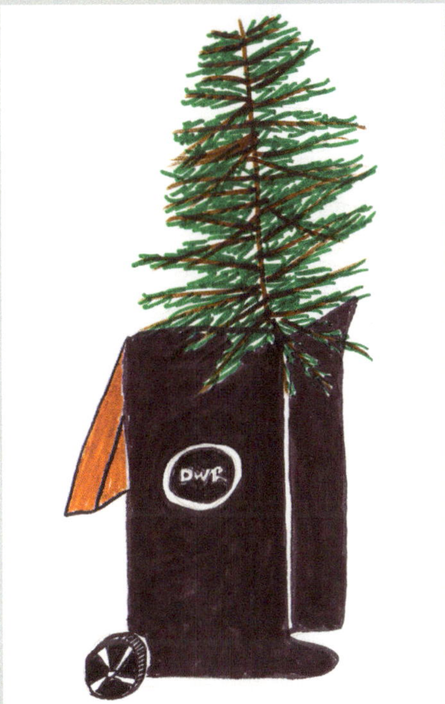
DWR
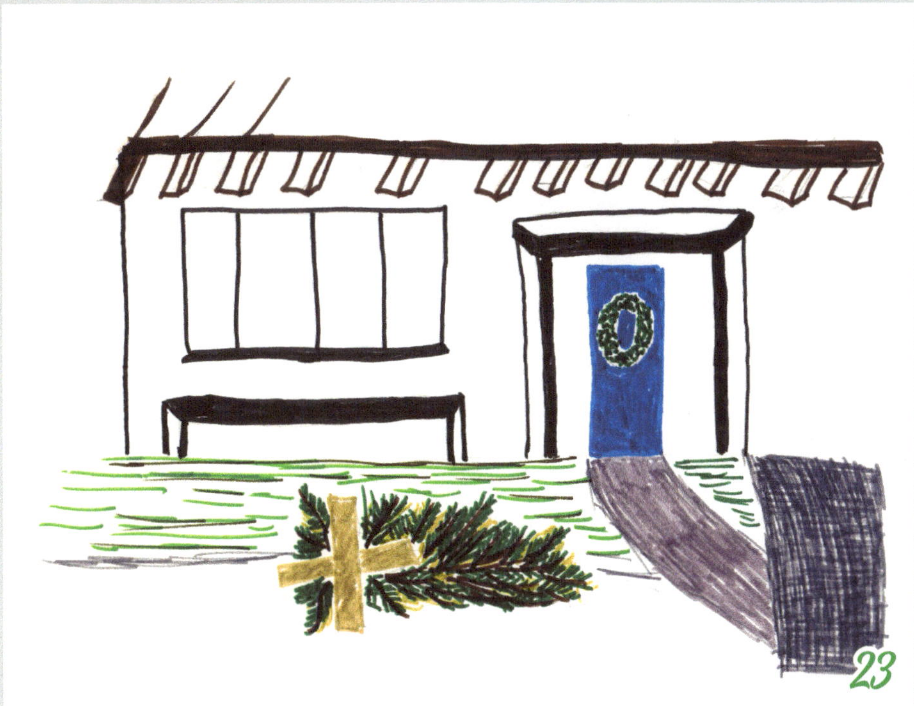

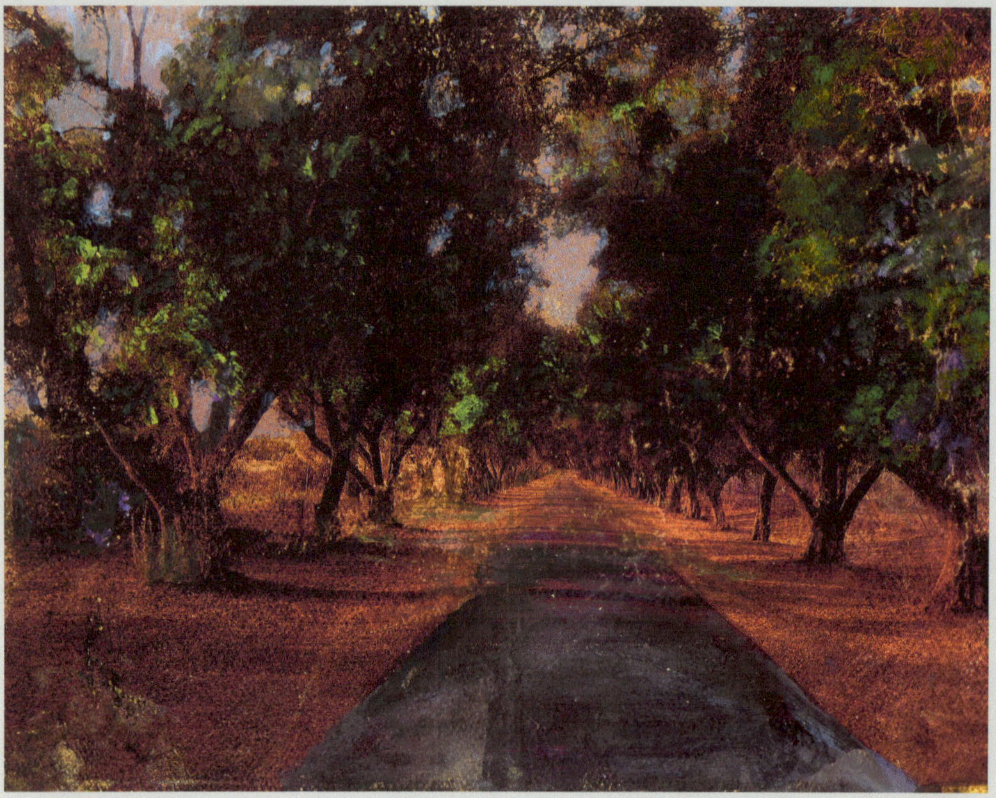

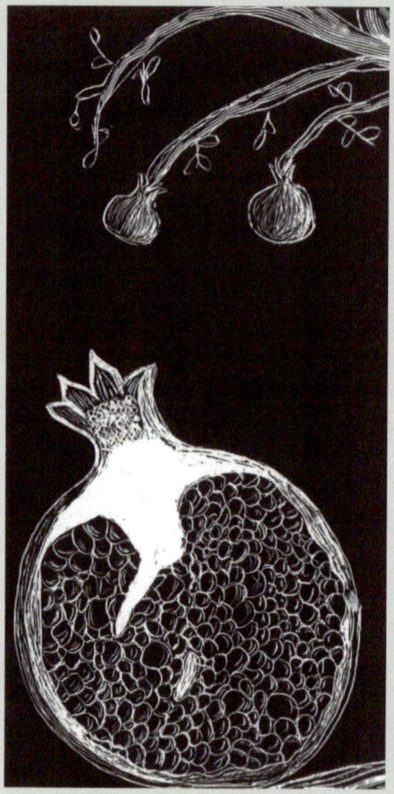

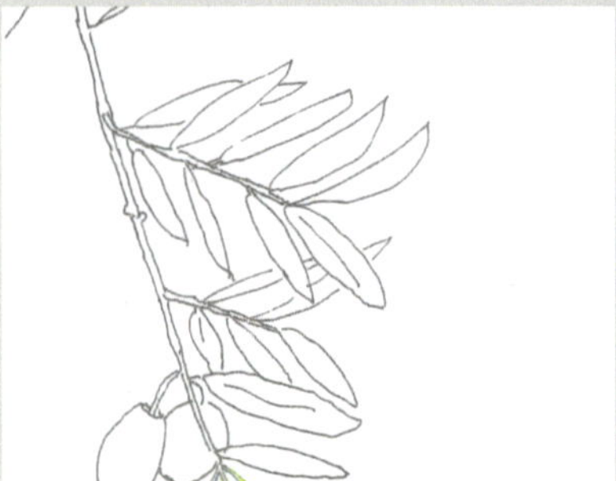

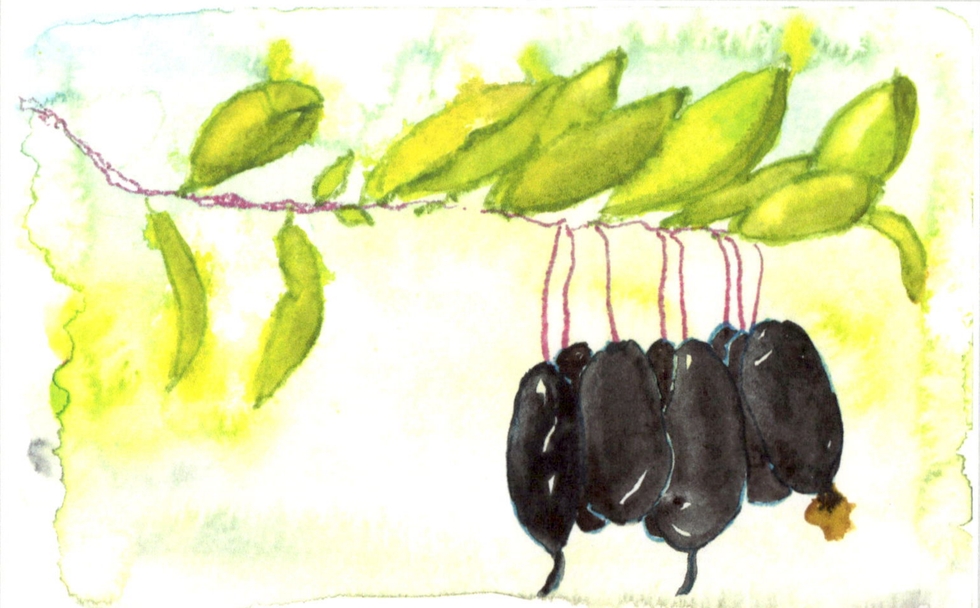

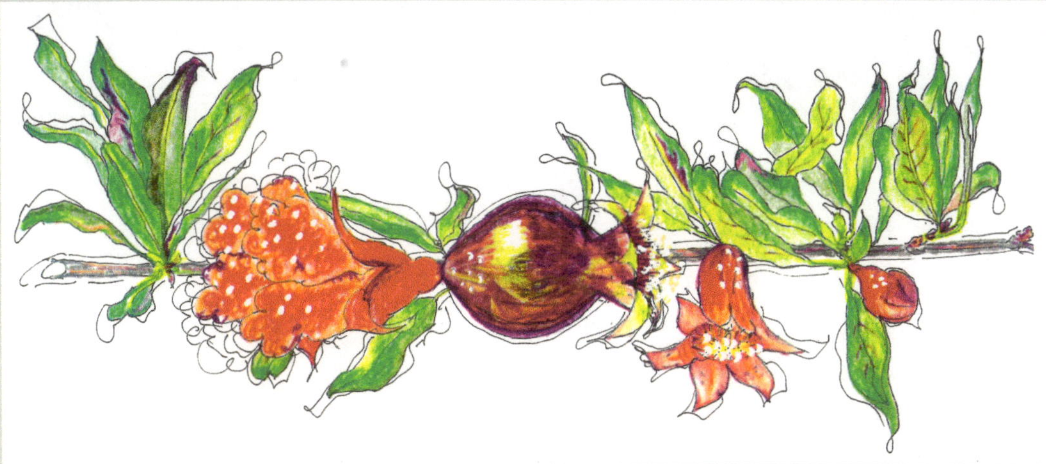

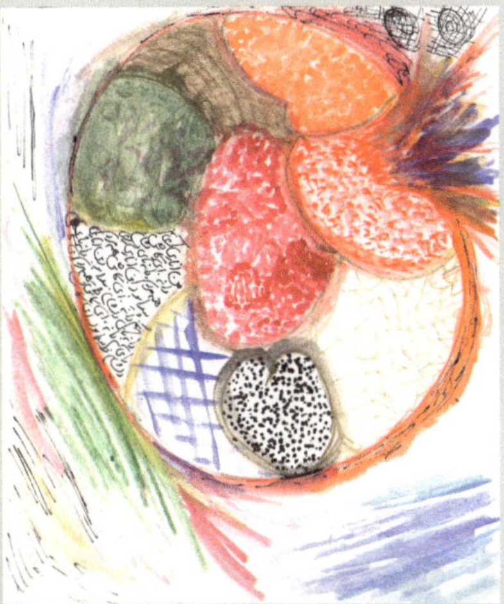

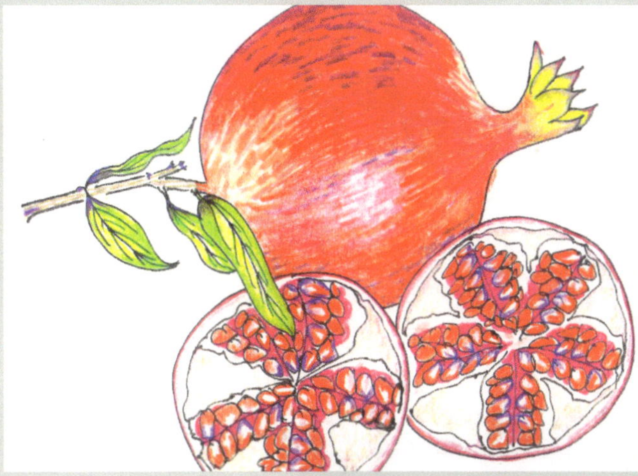

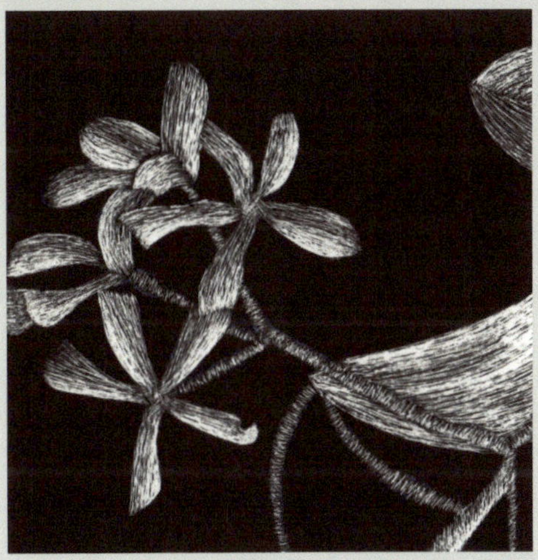

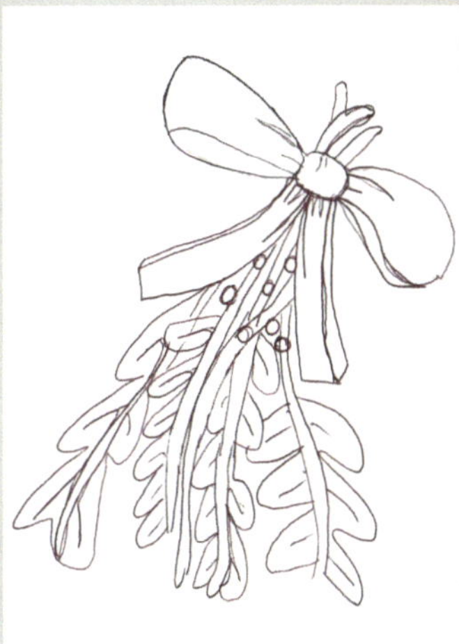

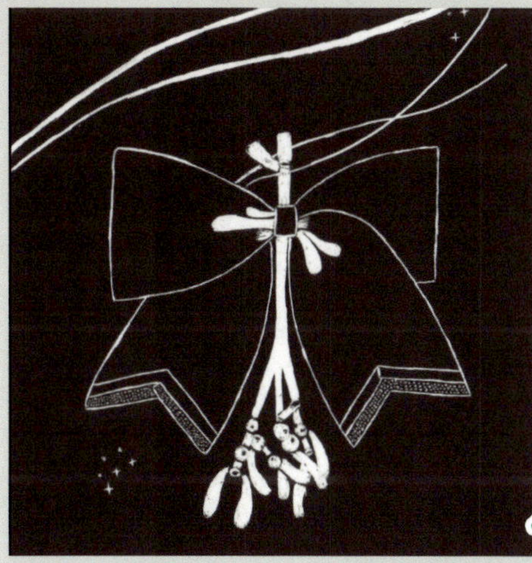

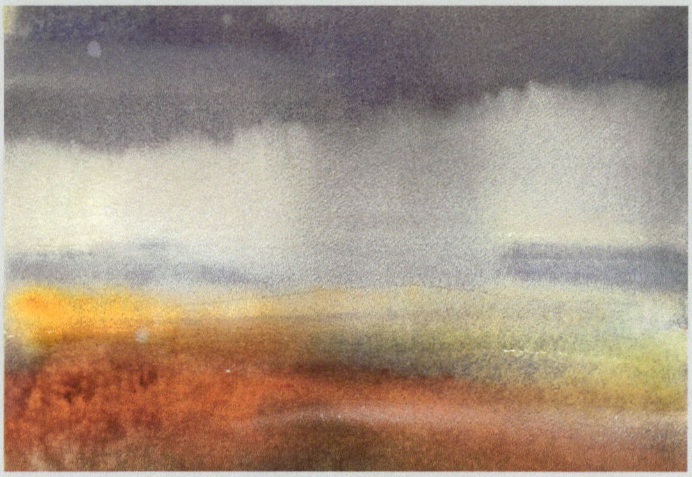
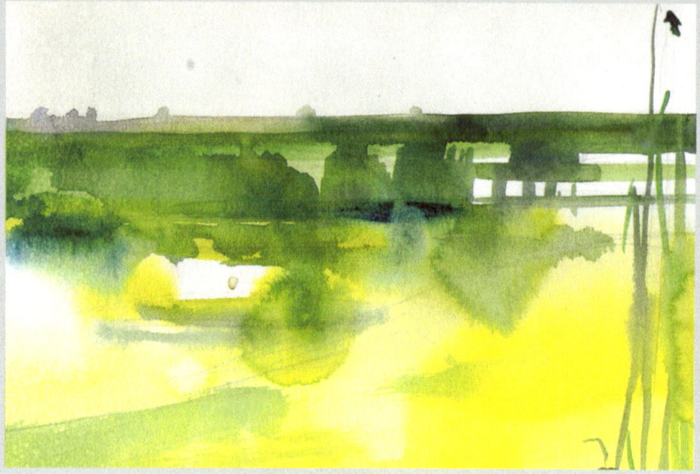
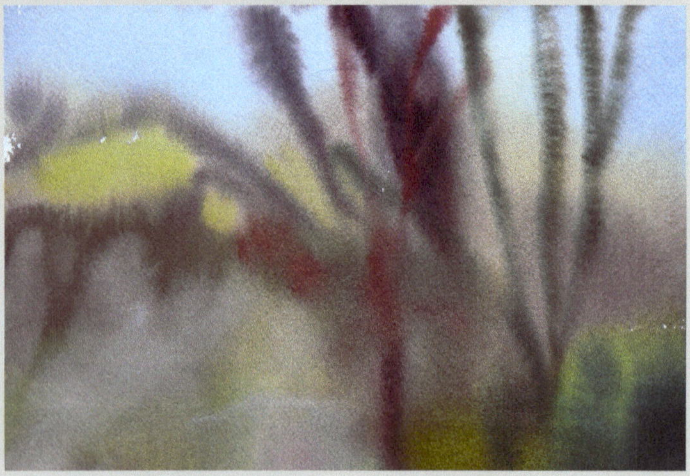
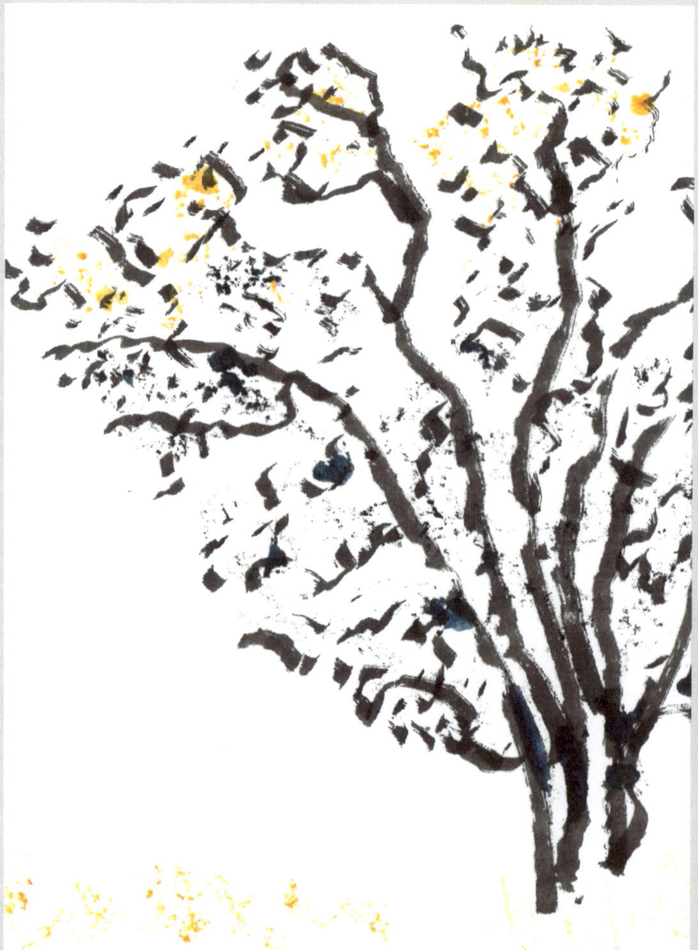
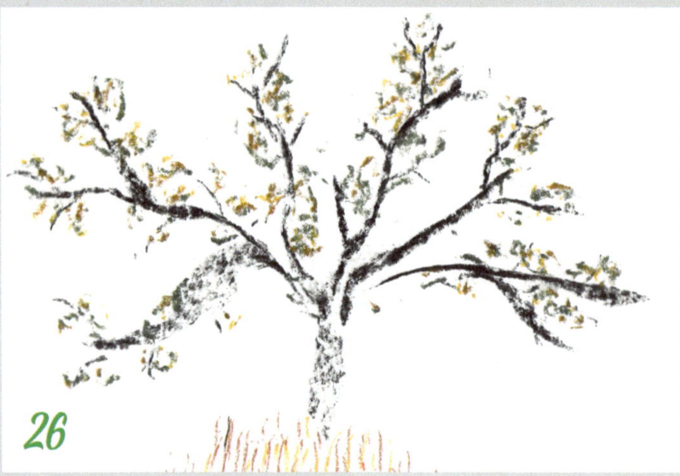

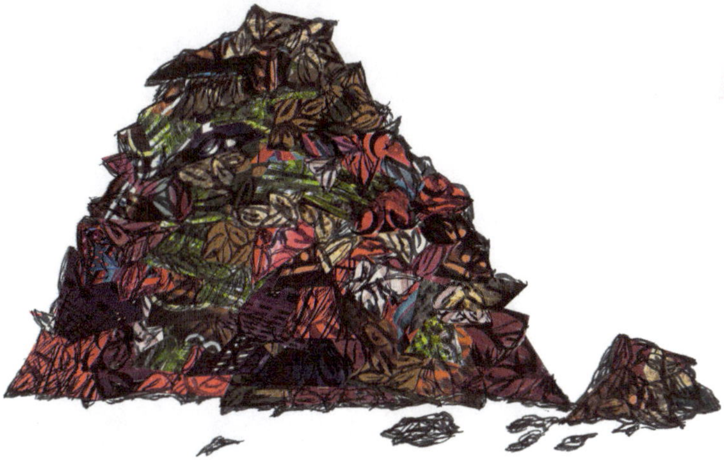
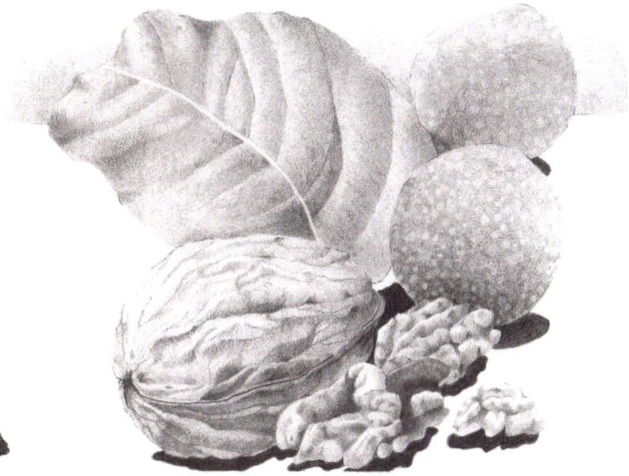
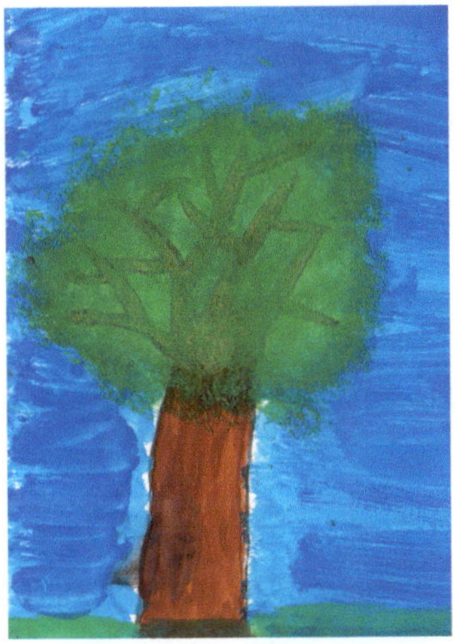
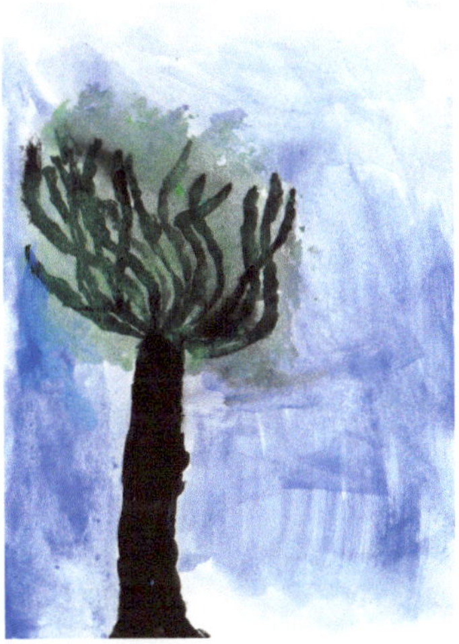
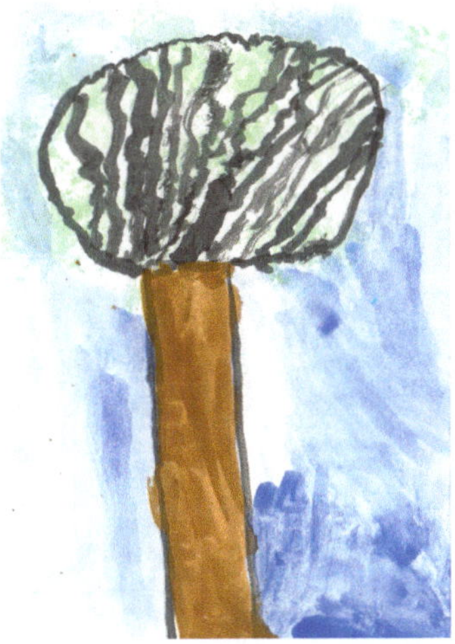
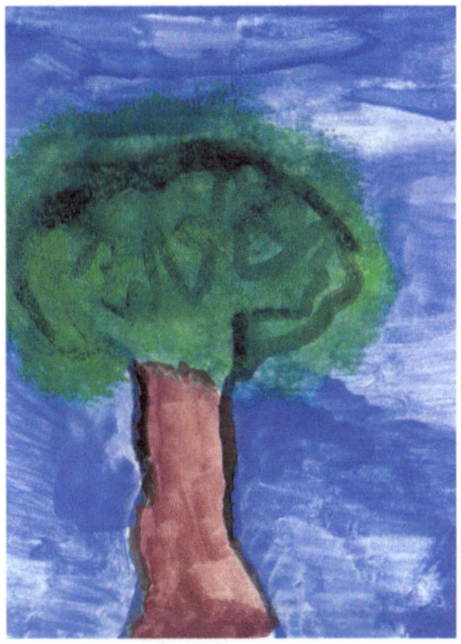
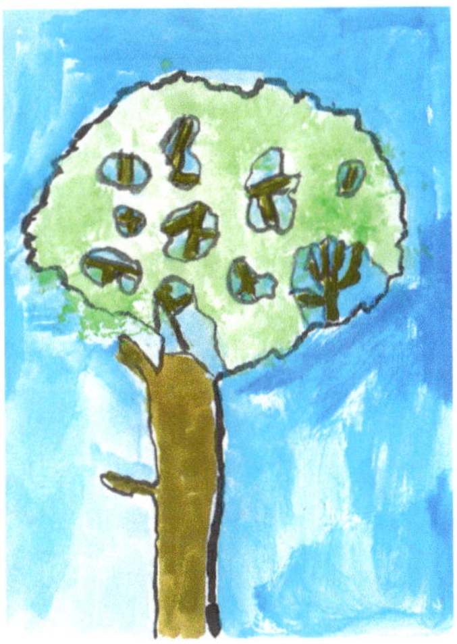
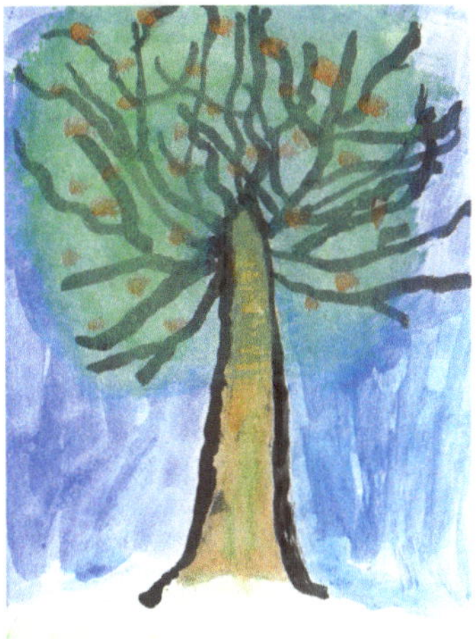
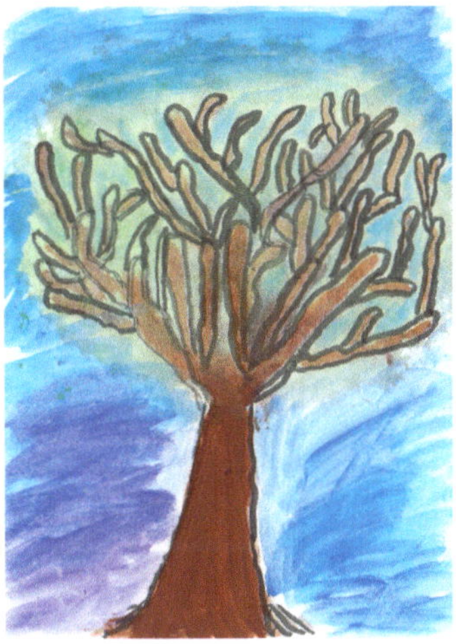
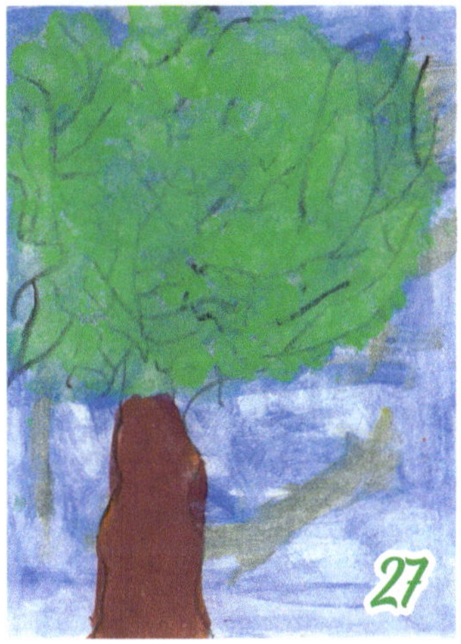

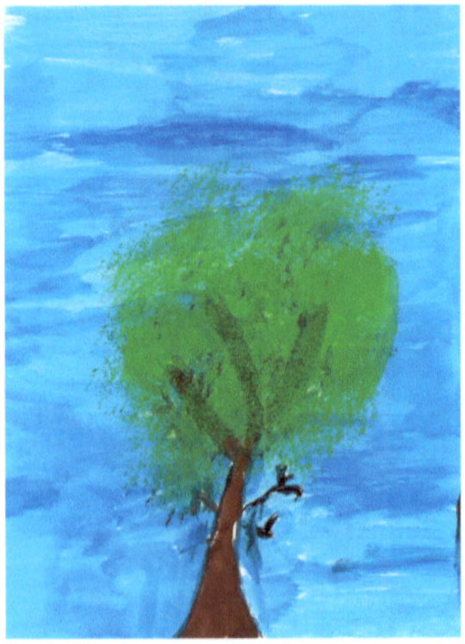
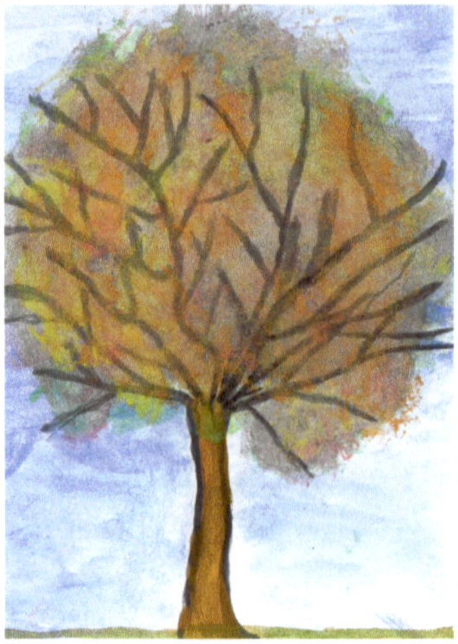
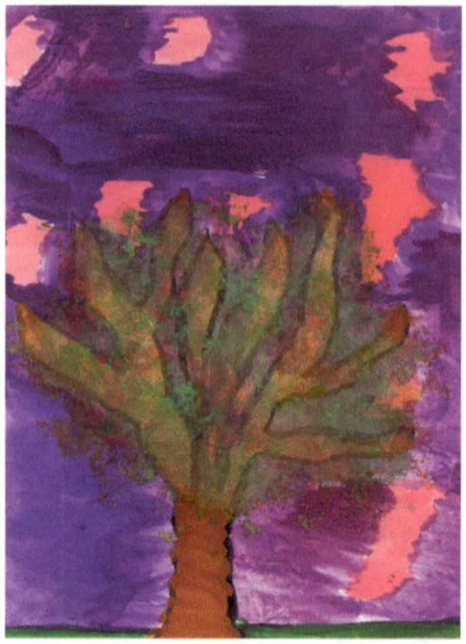
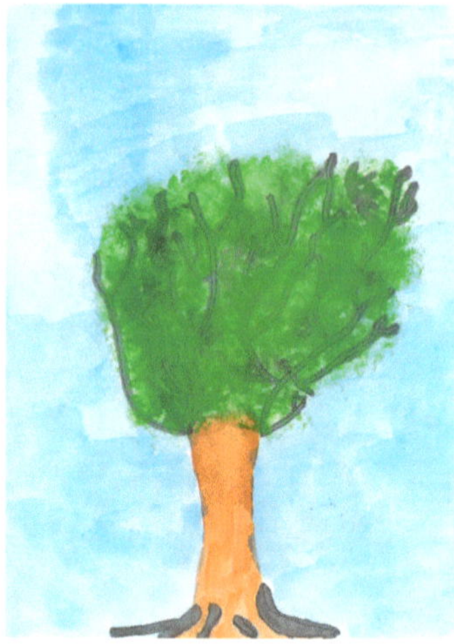
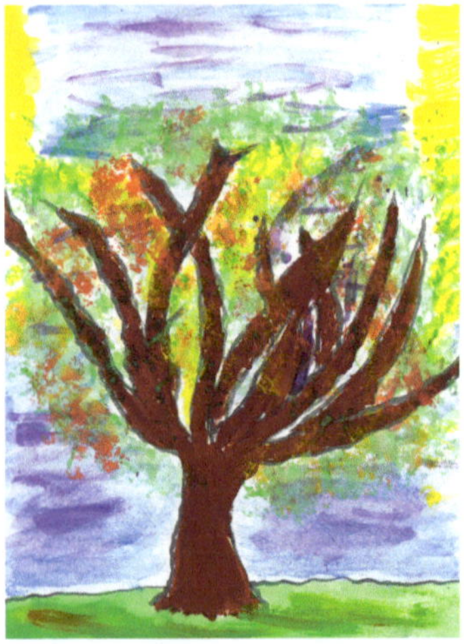
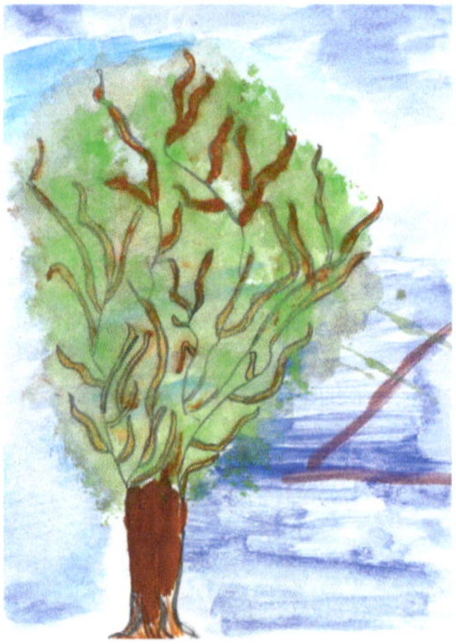
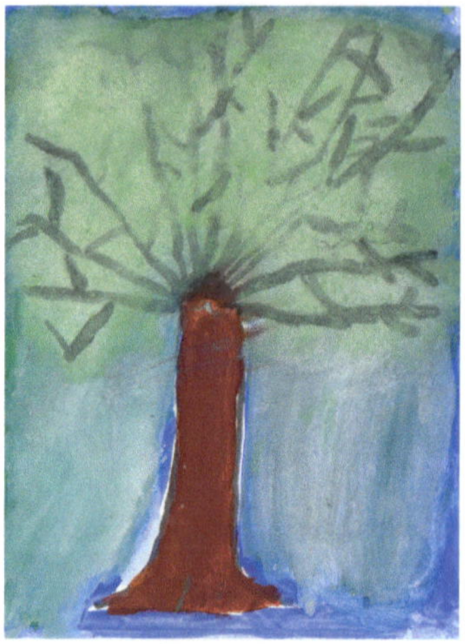
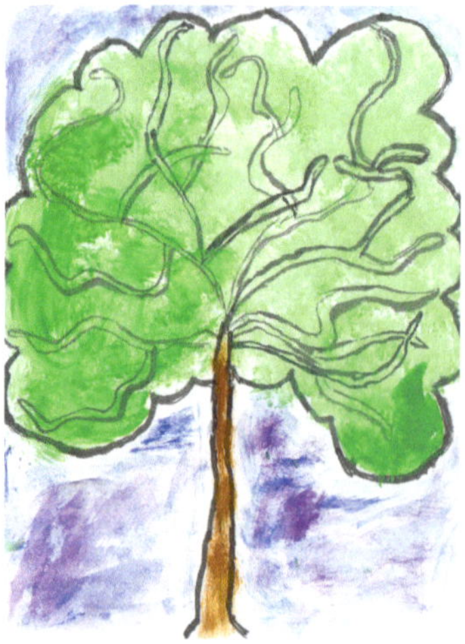
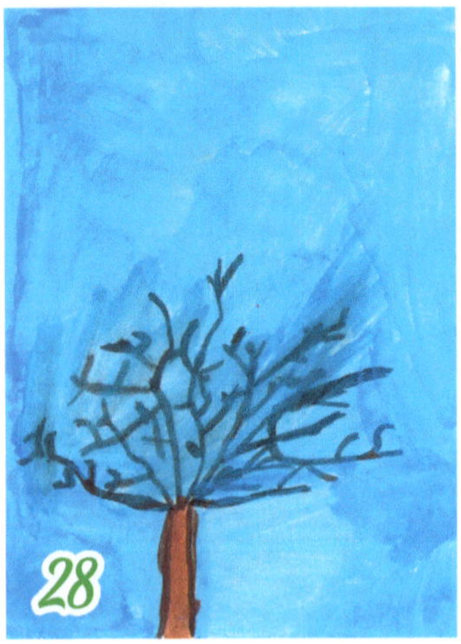
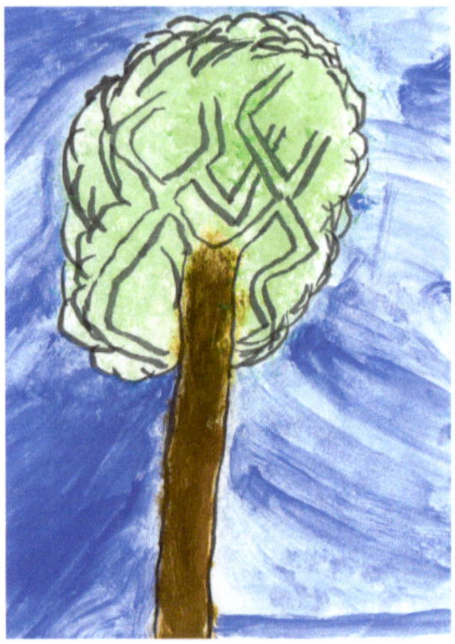
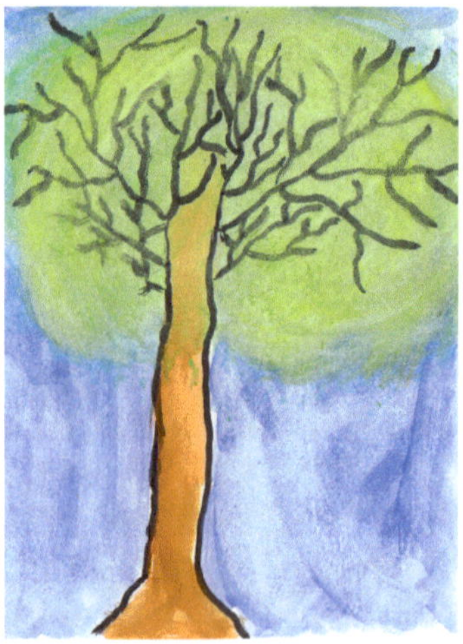
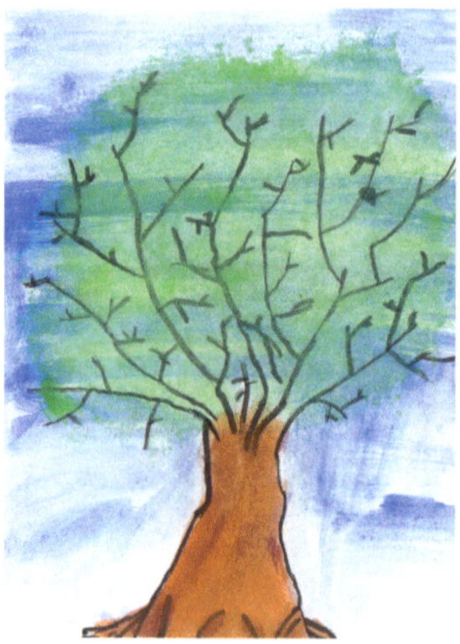

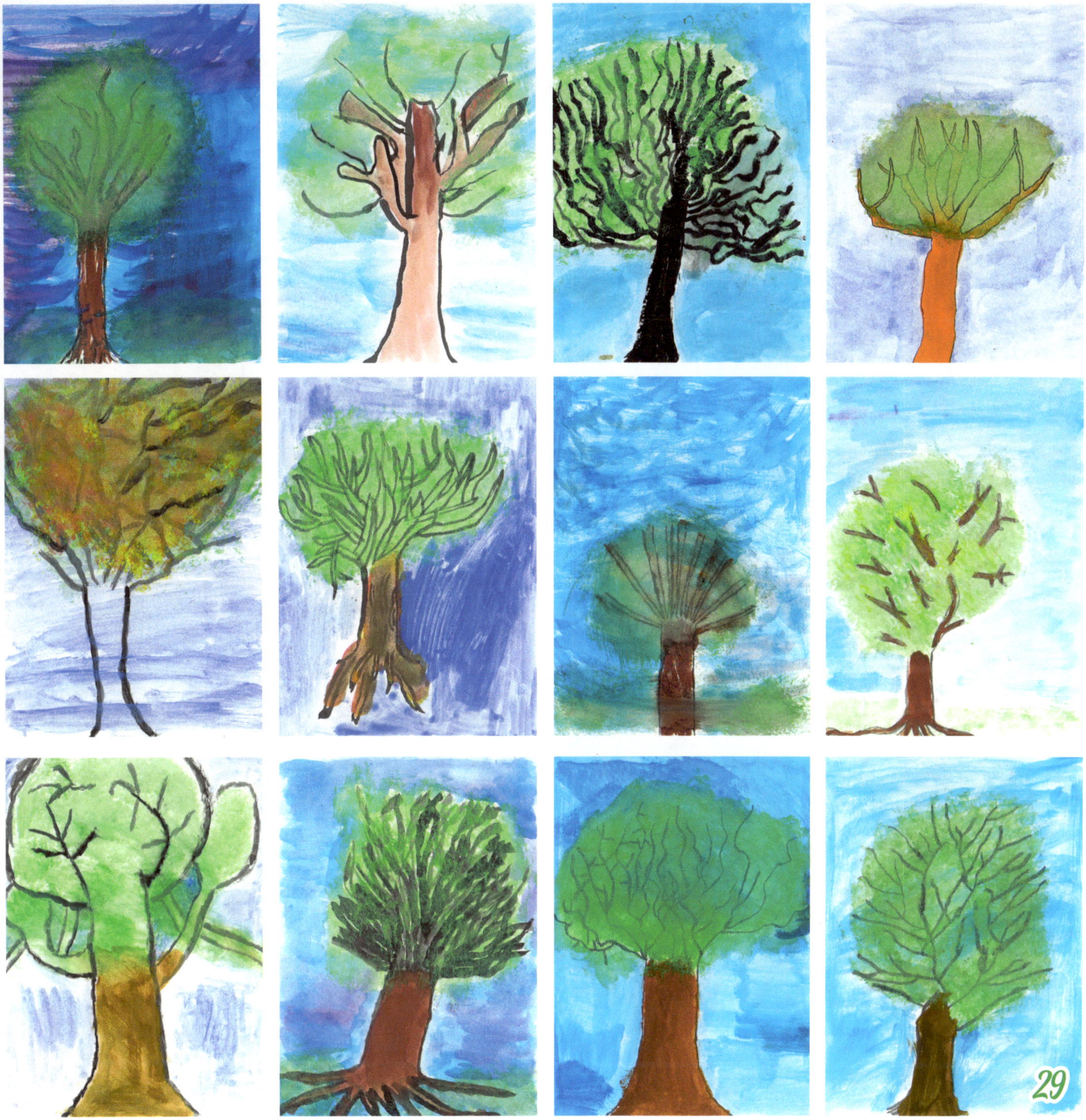

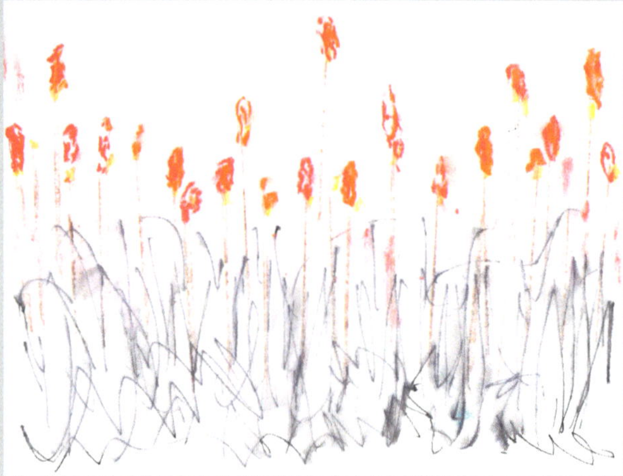

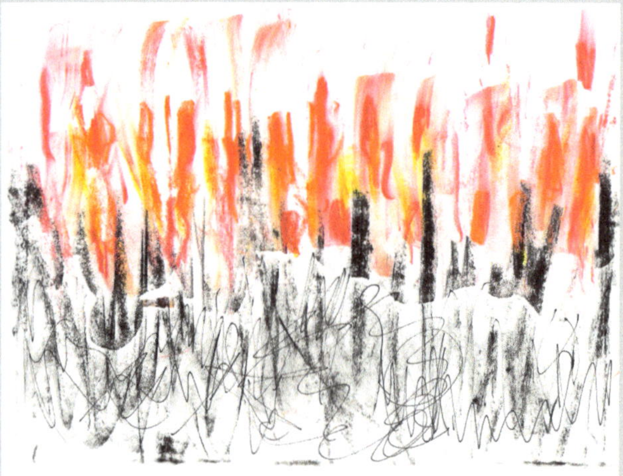

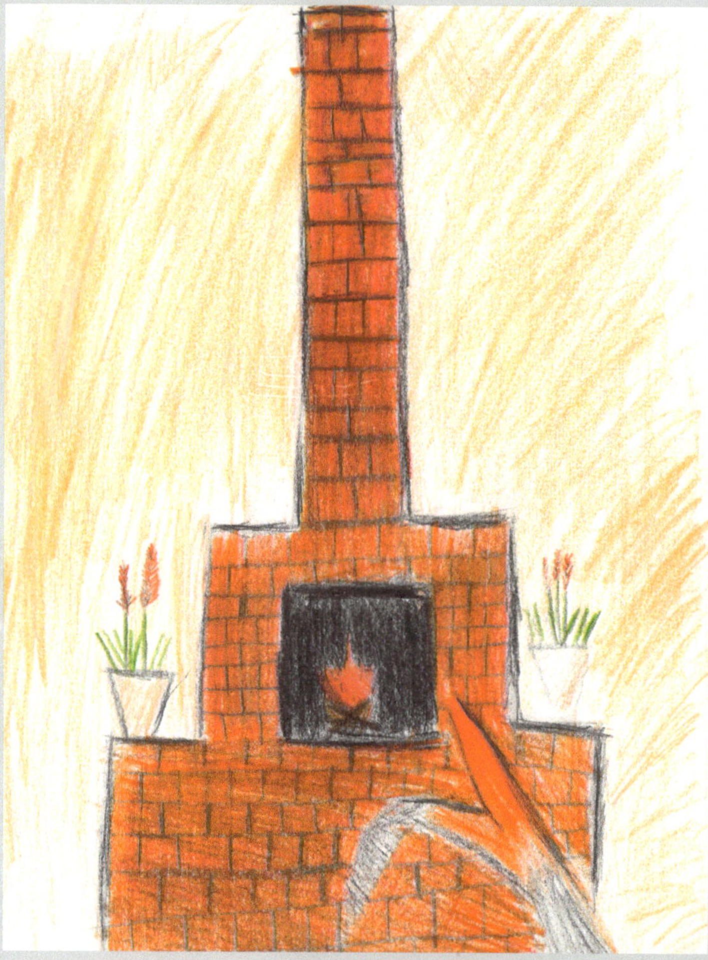

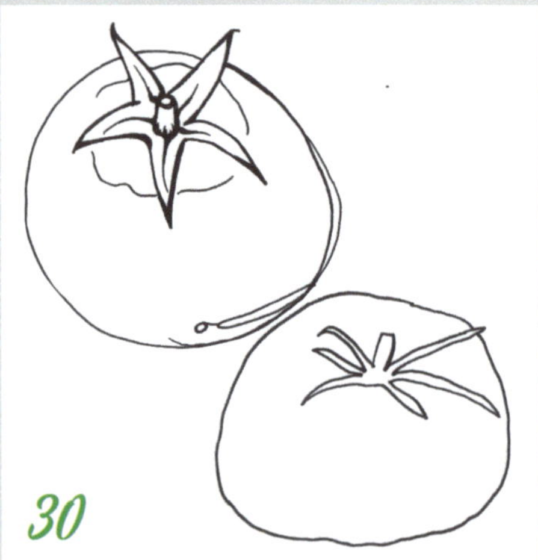

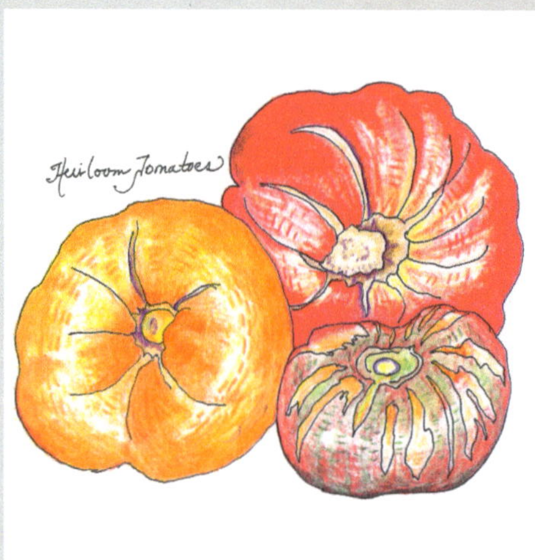
Heirloom Tomatoes

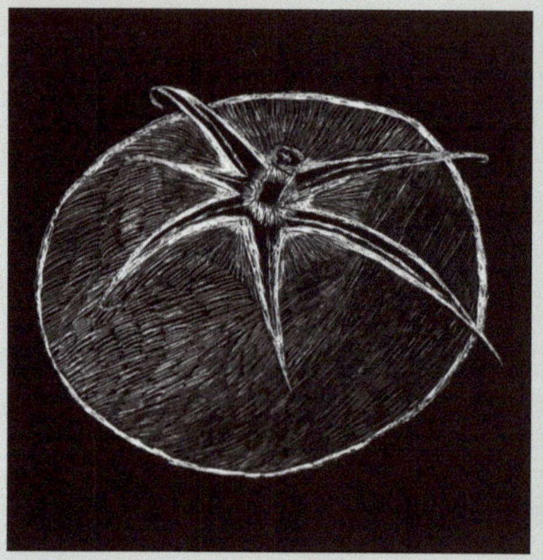

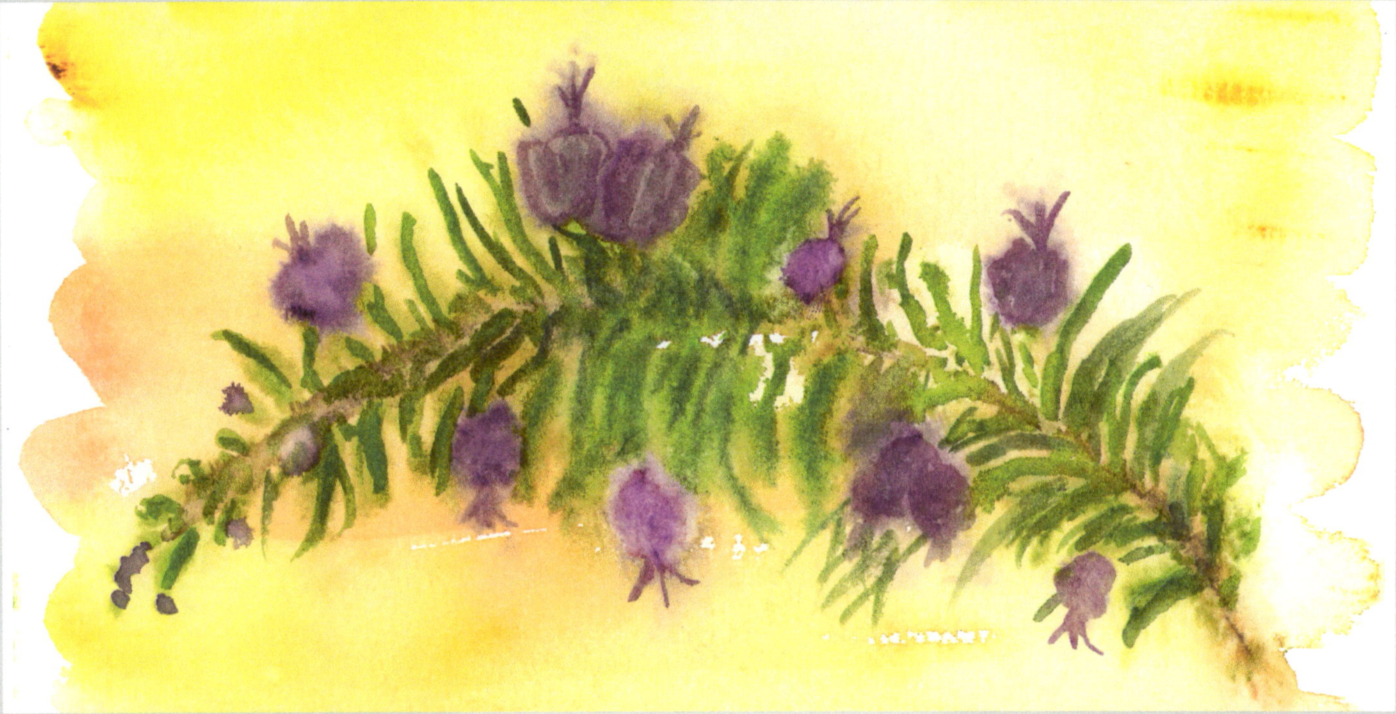

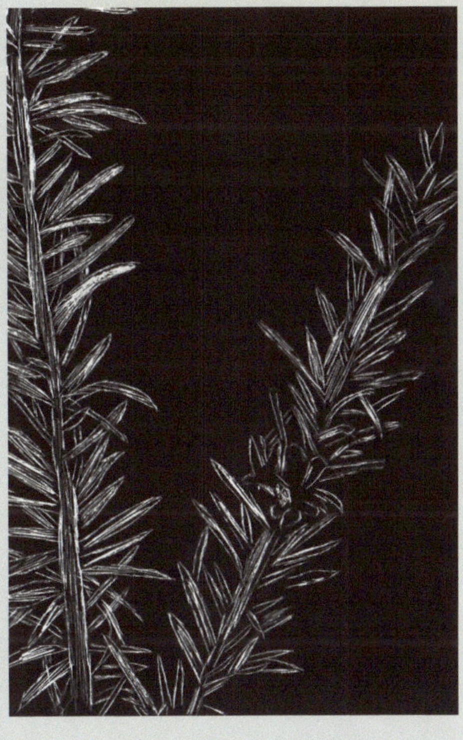

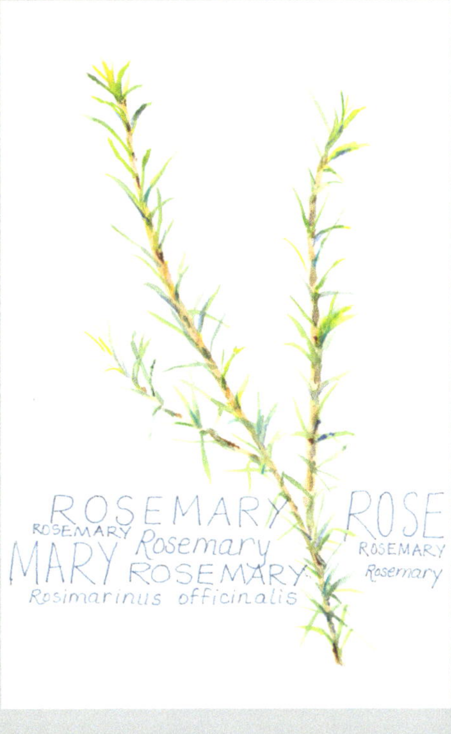
ROSEMARY
ROSE
ROSEMARY
Rosemary
ROSEMARY
MARY
ROSEMARY
Rosemary
Rosimarinus officinalis

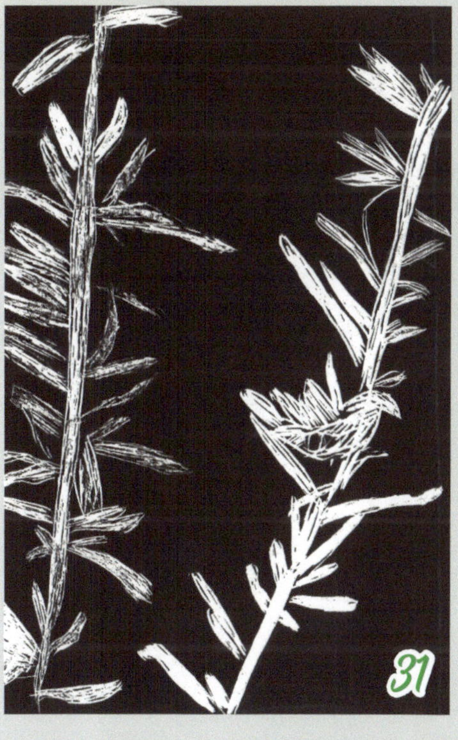

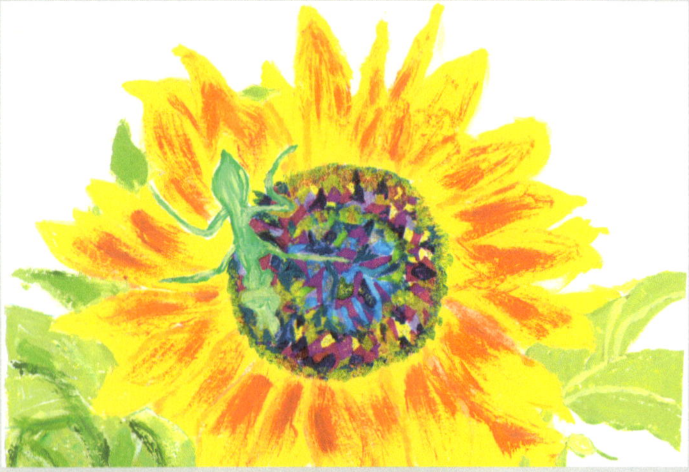

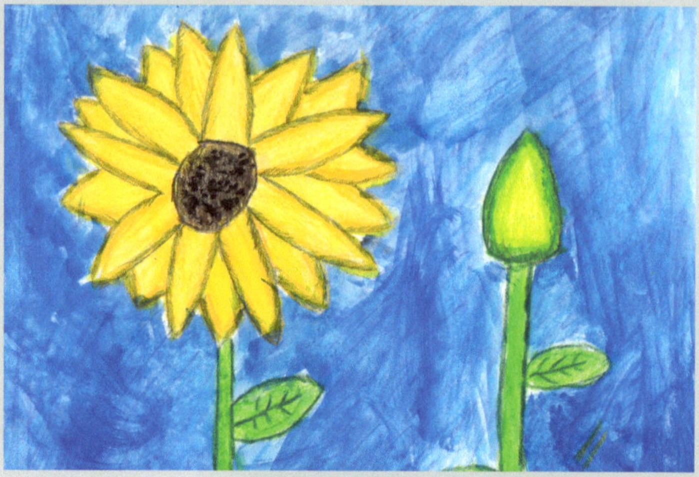

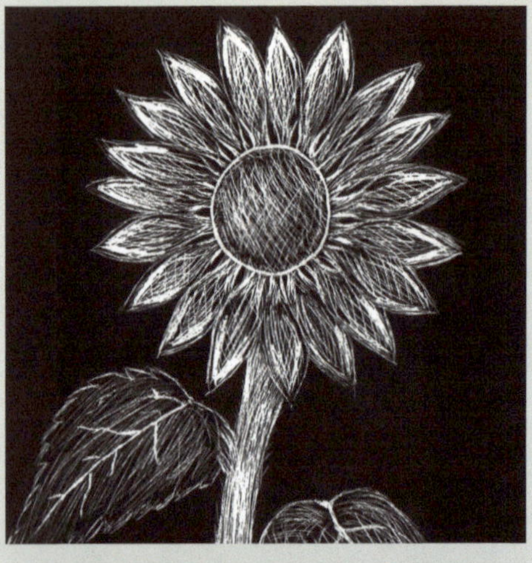

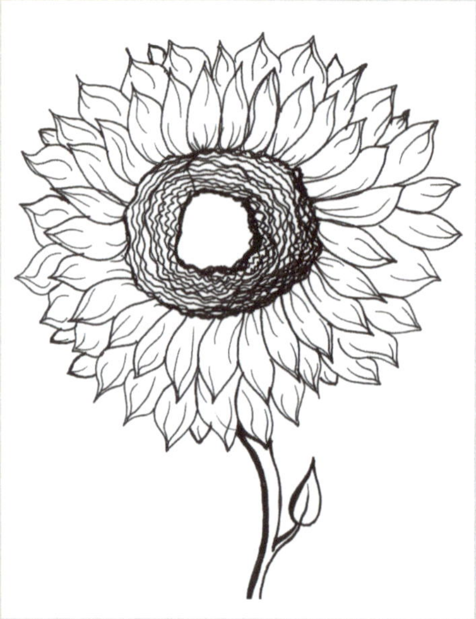

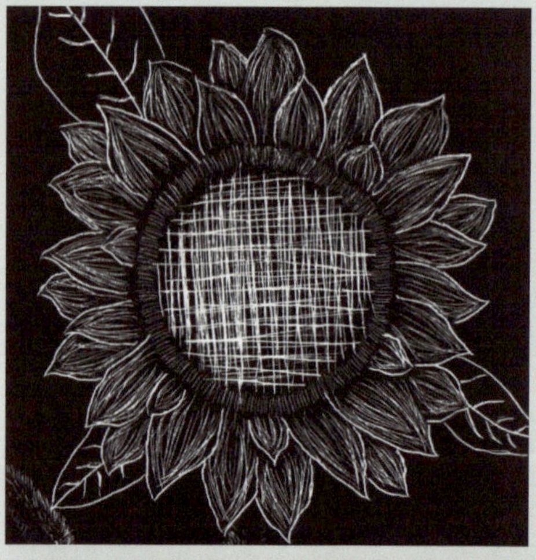

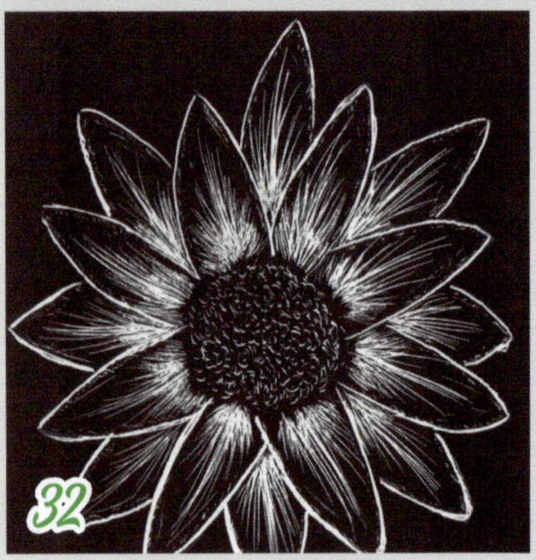

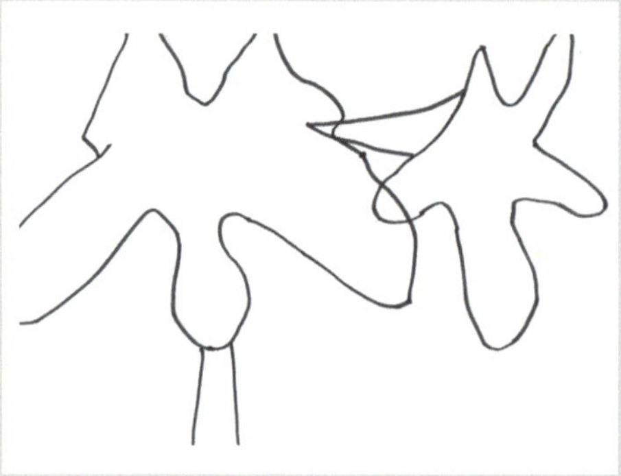

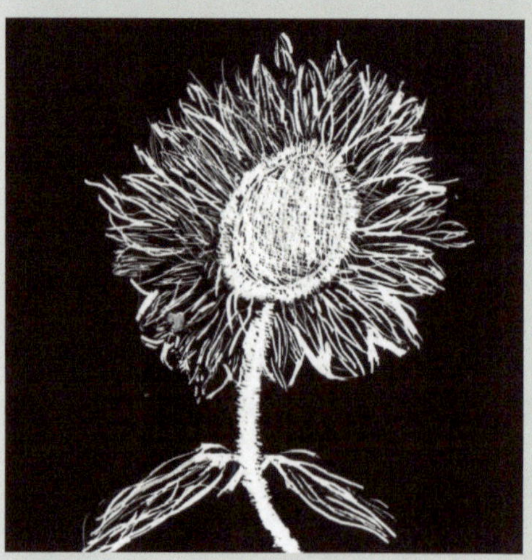

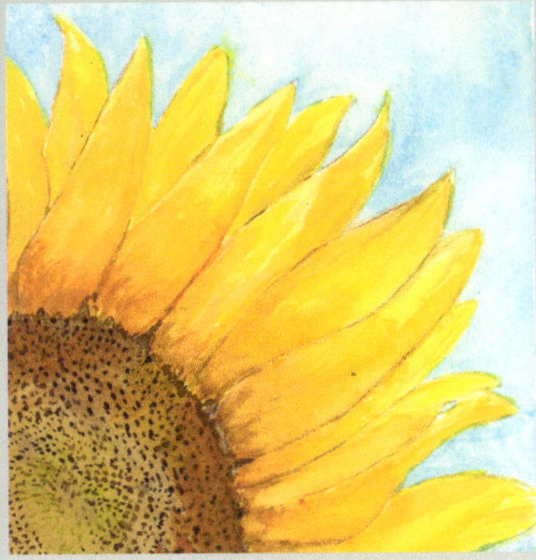
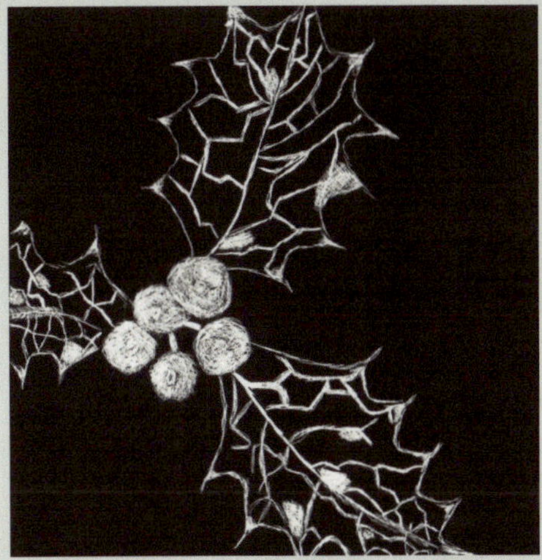
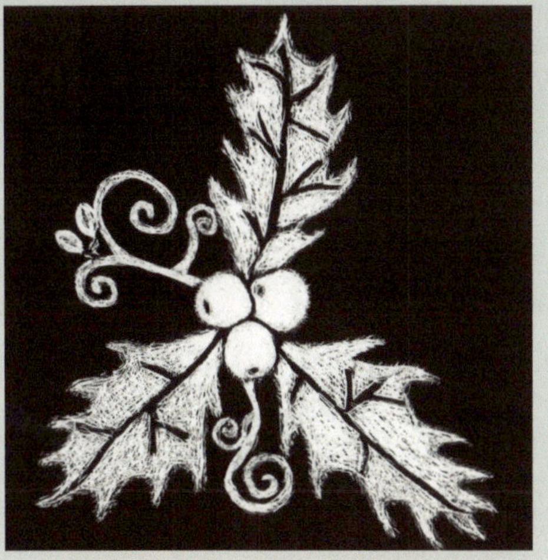
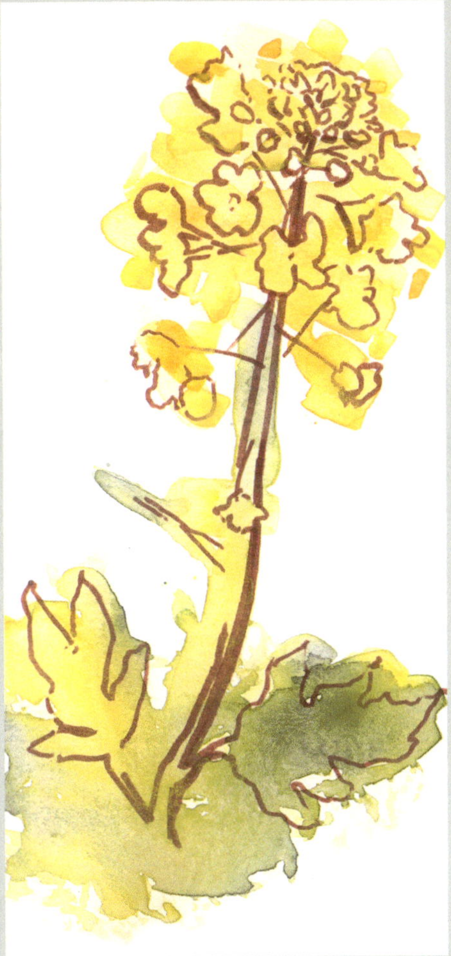
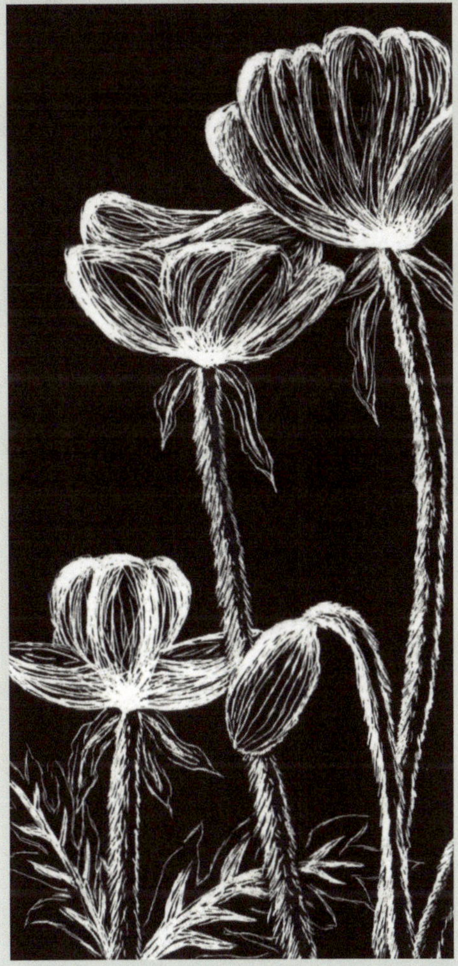
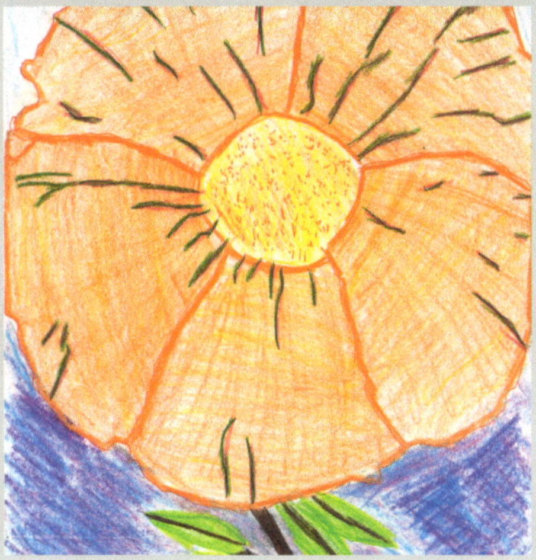
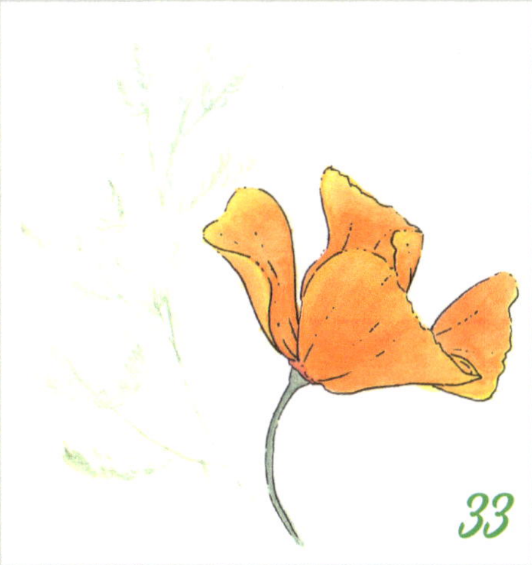

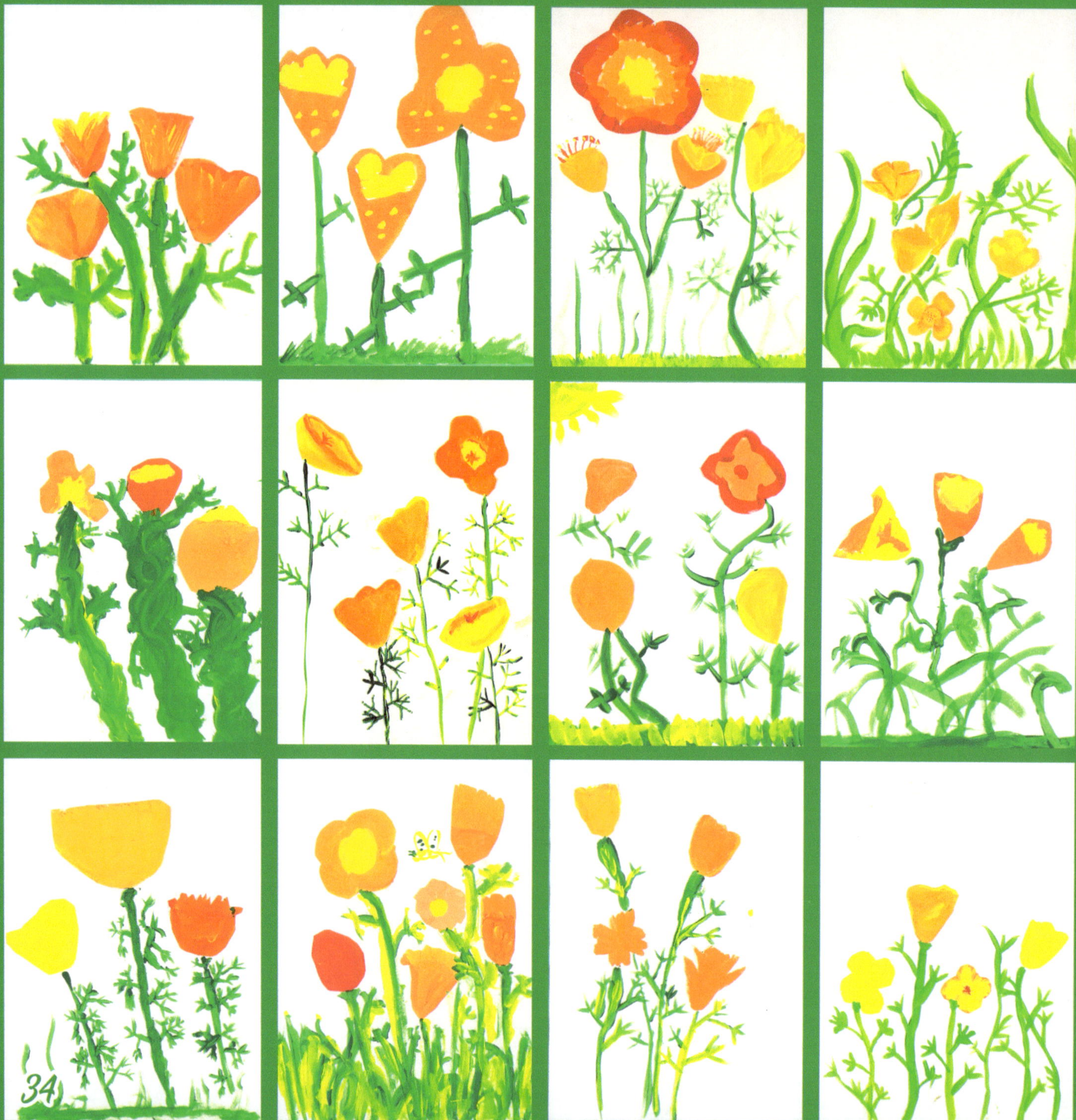

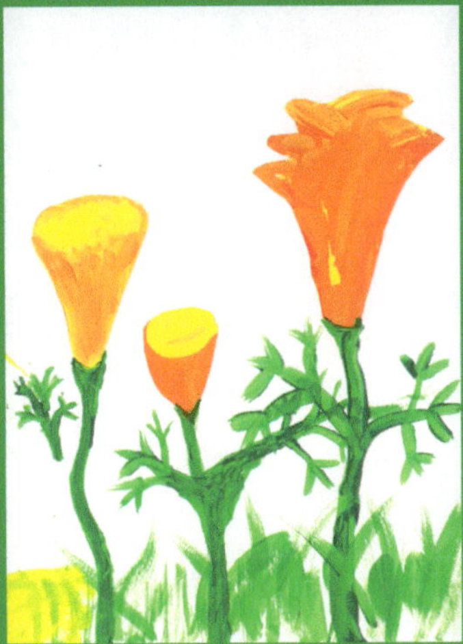
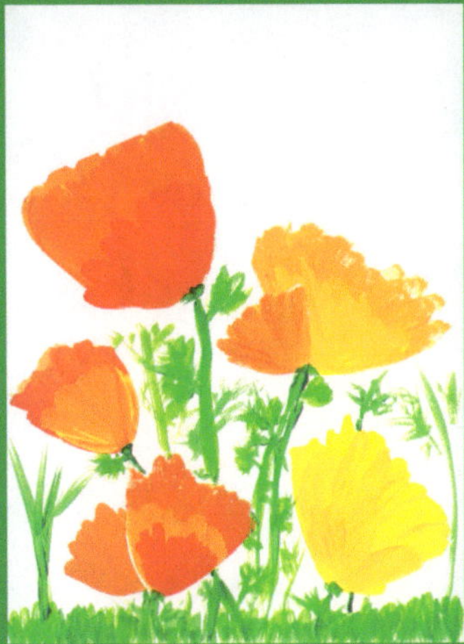
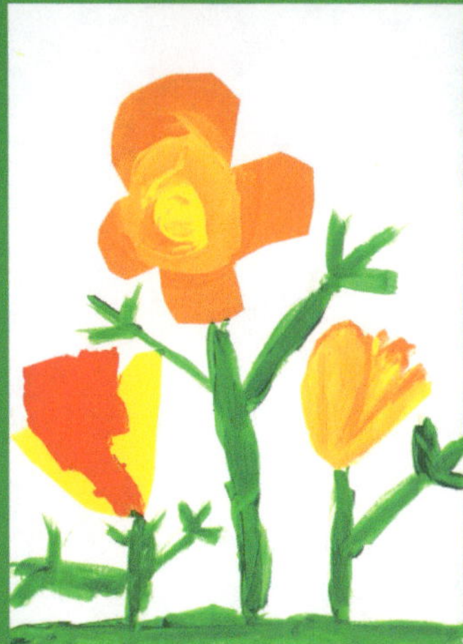
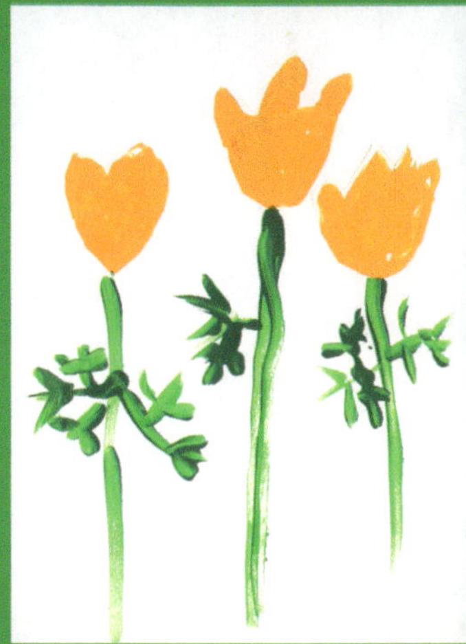
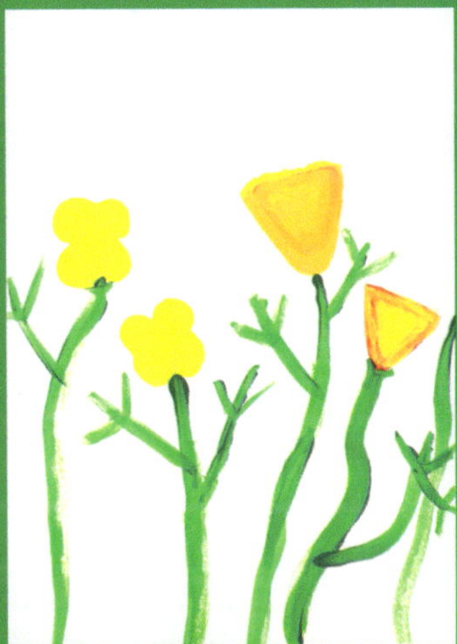
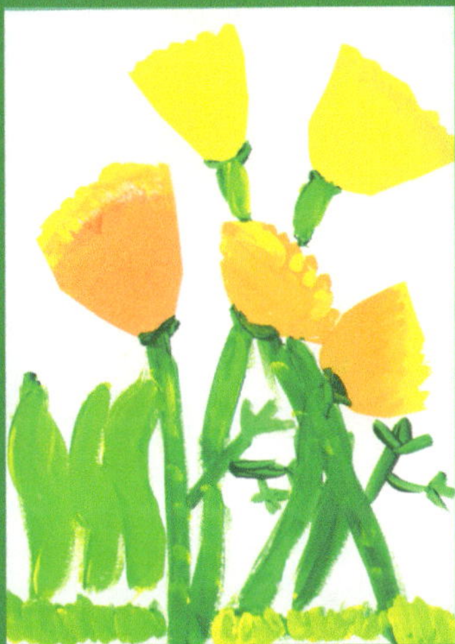
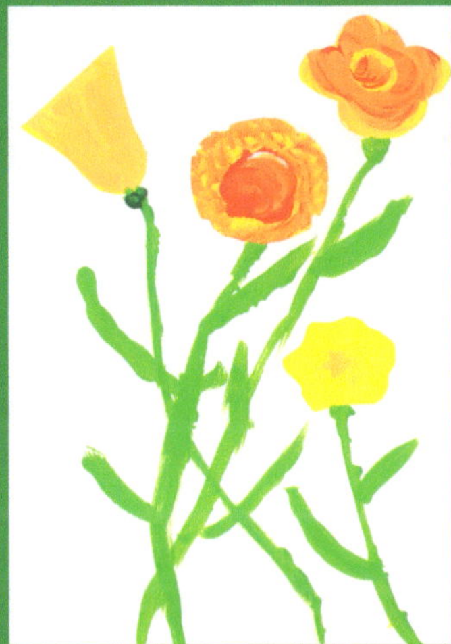
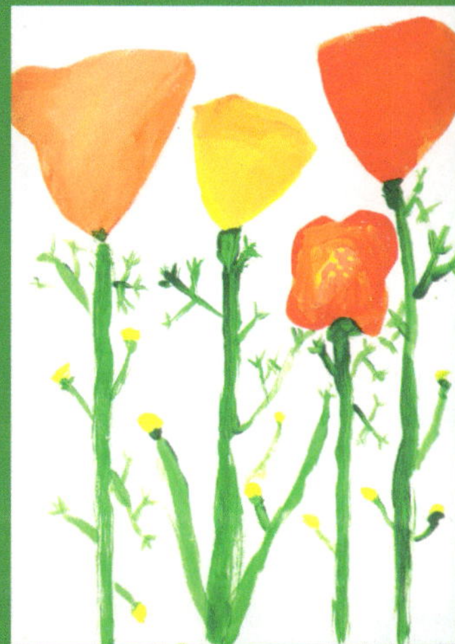
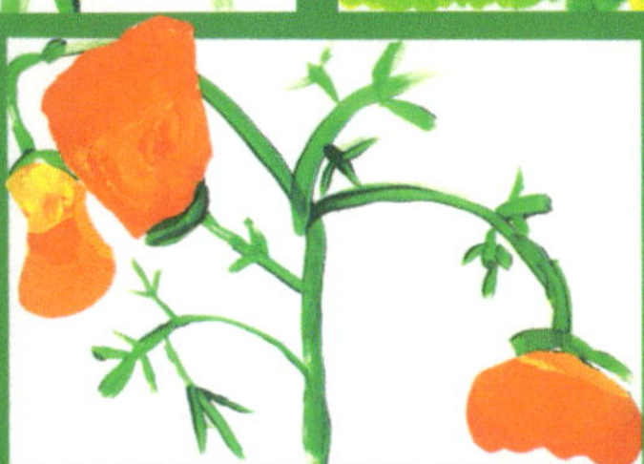
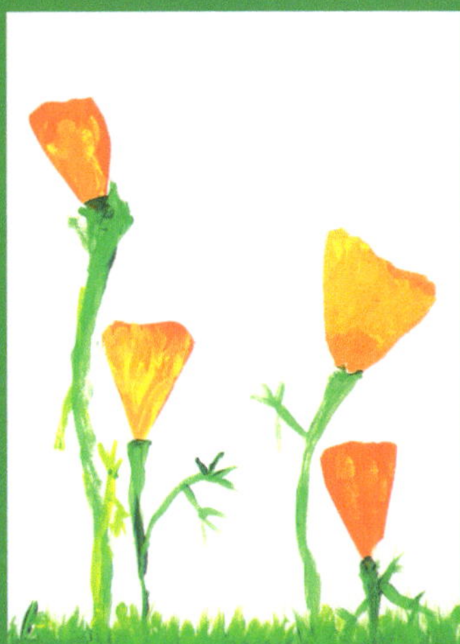
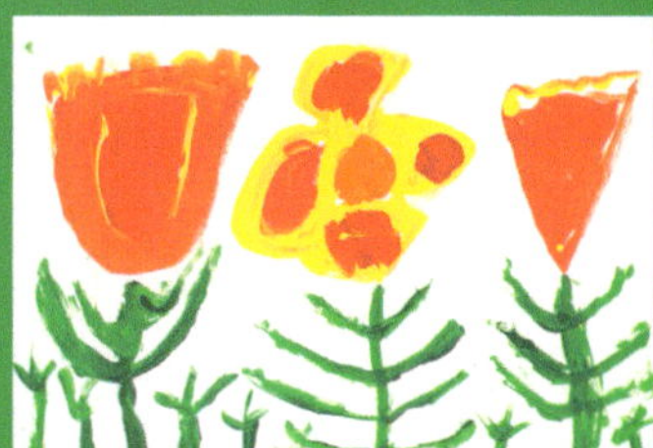
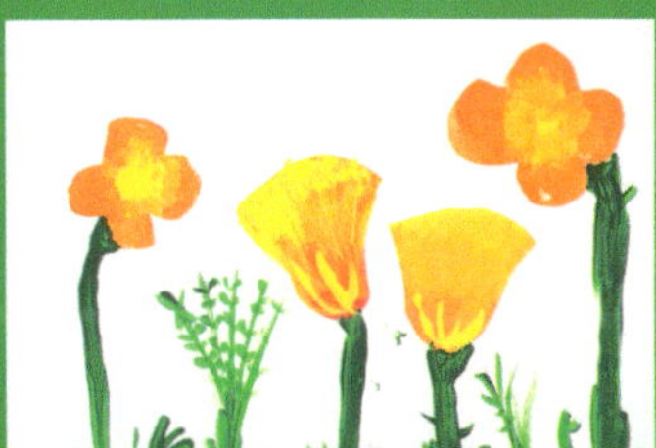
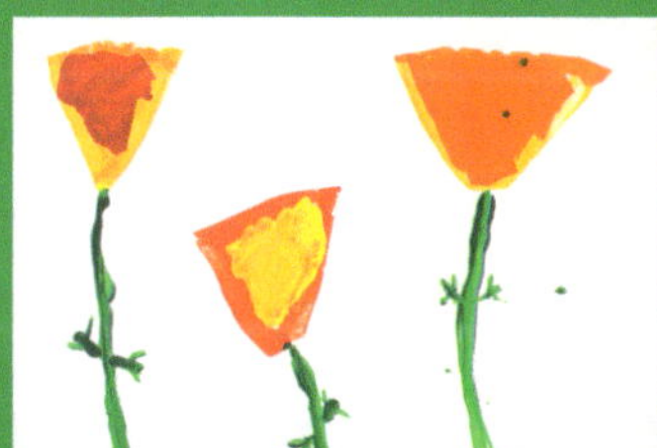

COYOTE • DOG • HORSE • OTTER • OPOSSUM • RABBIT • RACCO

AT • SQUIRREL • INSECTS • BEE • BLACK WIDOW SPIDER • BUTTER

OCKROACH • CRANE FLY • DEERFLY • DRAGONFLY • FLY • MOSQU

OTH • PRAYING MANTIS • FANTASY • BIRDS • BLACKBIRD • CANADA G

CHICKEN • CROW • DOVE • DUCK • EAGLE • EGRET • FINCH • FLICKE

AWK • HERON • HUMMINGBIRD • MAGPIE • MOCKINGBIRD • OWL • SC

Y • STARLING • SWALLOW • TANAGER • TURKEY • TURKEY VULTU

ANDSCAPE • CLOUDS • DELTA BREEZE • FLATNESS • FOG • NORTH WI

UTAH CREEK • SUNSET • PLANTS • ACORN • BERMUDA GRASS • ELDERBE

BUCKEYE • BULBINE • CEANOTHUS • CENTURY PLANT • CHINESE PISTA

CHINESE TALLOW • COTTONWOOD • DAFFODIL • DATURA • DISCAR

HRISTMAS TREE • ELDERBERRY • FIG • FLOWERING QUINCE • FOX

RASS • GINKGO • HACKBERRY • HEUCHERA • IRIS • JASMINE • LEAF

LEMON • LUPINE • MILKWEED • MISTLETOE • MODESTO ASH • MUST

OLIVE TREE • ORNAMENTAL PEAR • PENSTEMON • PERSIMMON •

POMEGRANATE • POPPY • PURPLE NEEDLE GRASS • RED HOT POK

EDBUD • REDWOOD • ROSEMARY • SALVIA • SOURGRASS • SUNFLOW

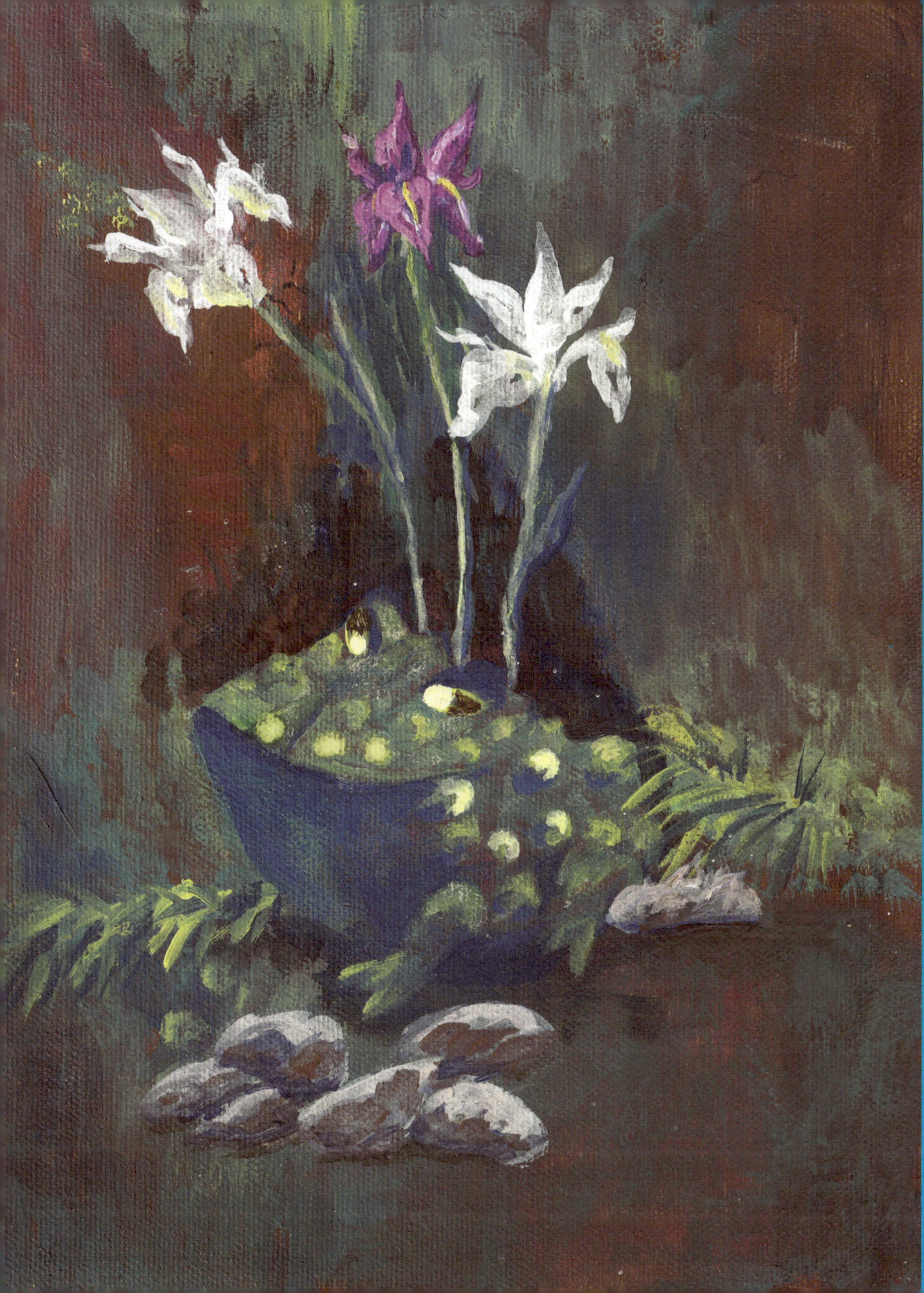

FISH & REPTILES

The following artists created the artwork in this section. The list is alphabetical by category and does not necessarily represent the order nor page number the artwork appears; please refer to the page numbers next to the category to locate where the art appears. Some teachers/schools did a class project (CP) for one category; all the artists involved in that project are listed in italics under the teacher and school name within the category. For other class projects the artist chose their category; their names appear under that category with a footnote indicating their teacher/school. If there is a number after an artist's name they provided more than one piece of art for that category.

FROG

(pages 37 and 40)
Chappuie, Lara - CP[1]
Sosnick, Galya

SALMON

(page 39)
Hicks, Bryan - CP[2]
Simpson, Jordan - CP[1]

TURTLE

(pages 40-41)
Carsey, Vanessa - CP[1]
M. Janie, - CP[2]
Nelson, Natalie (2)
Rice, Emily - CP[1]

[1]Student's Choice Class Project, Teacher: Melanie Bown, St. Francis High School • [2]Student's Choice Class Project, Teacher: Carrie Ceniseroz, St. James School

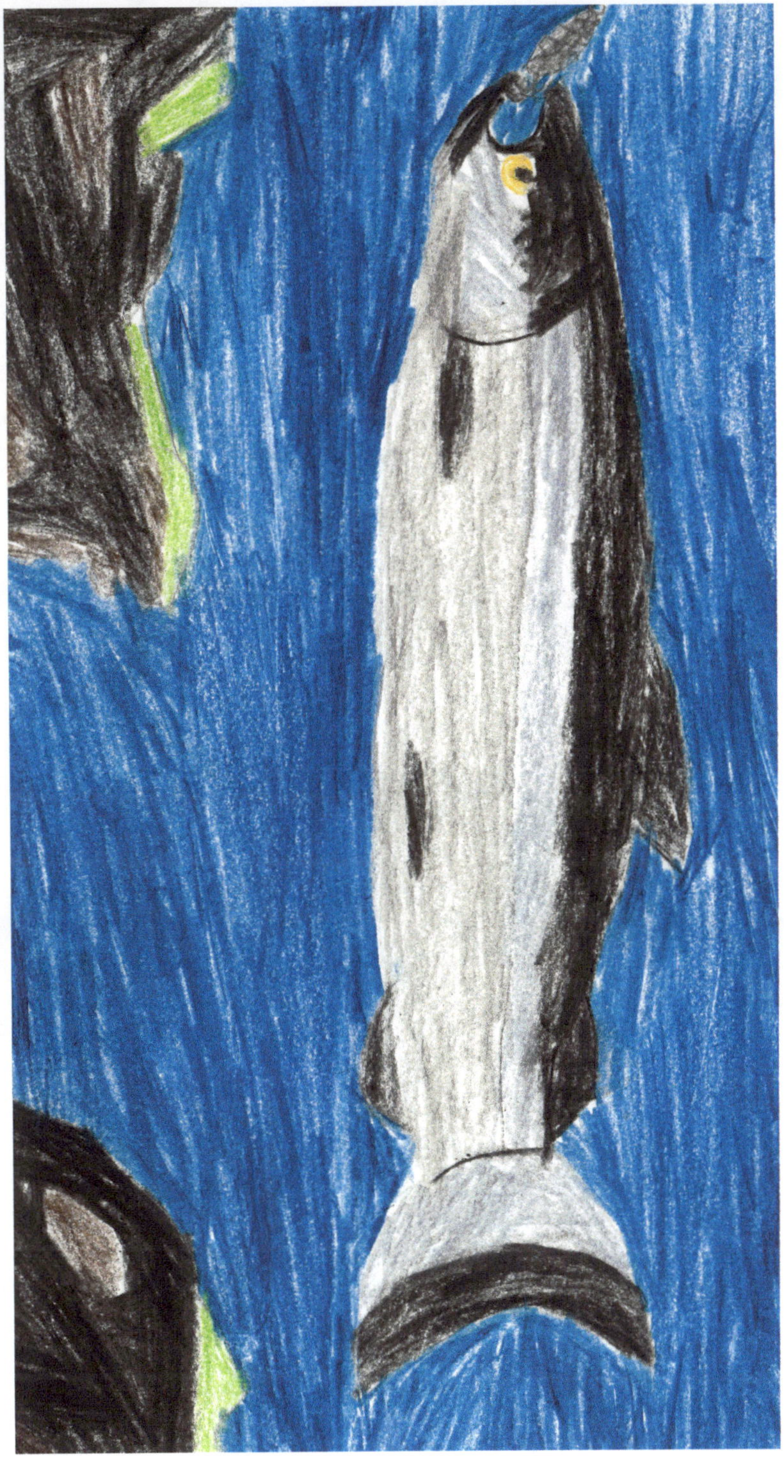

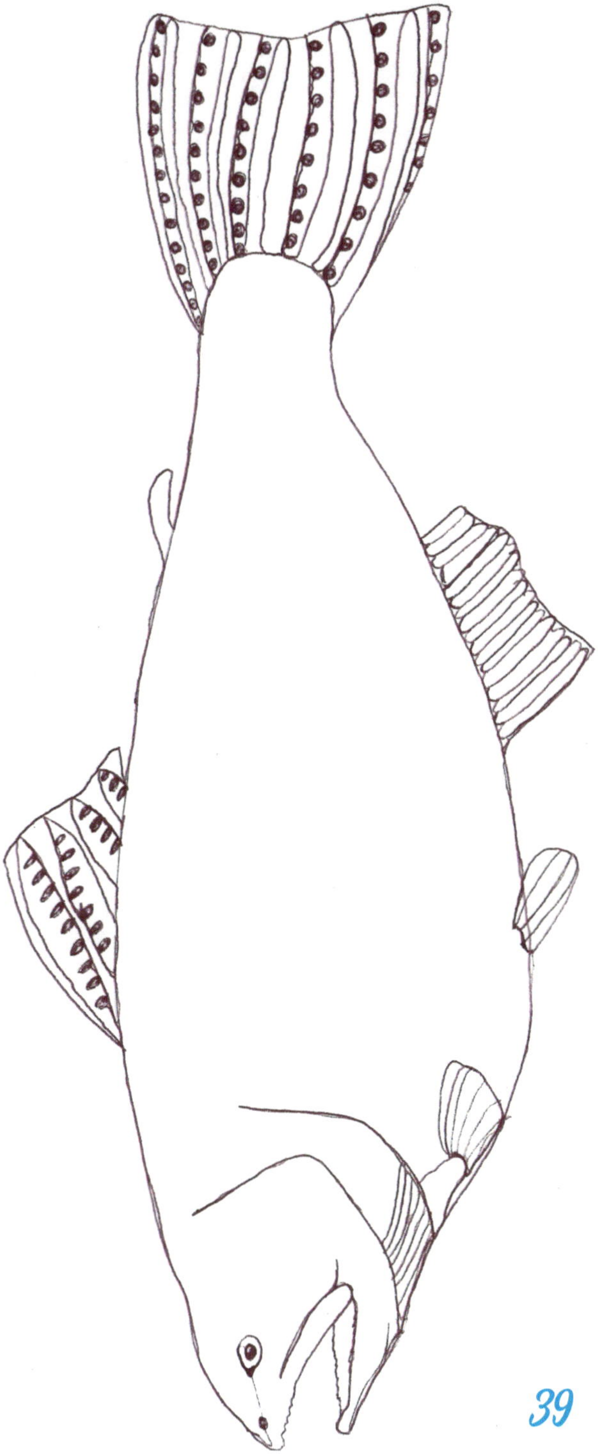

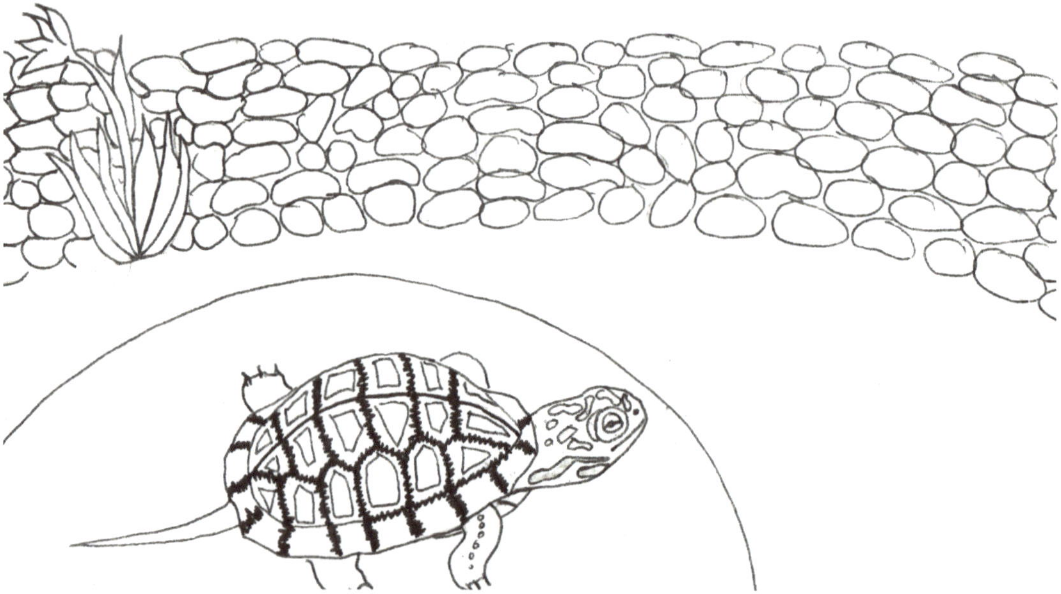

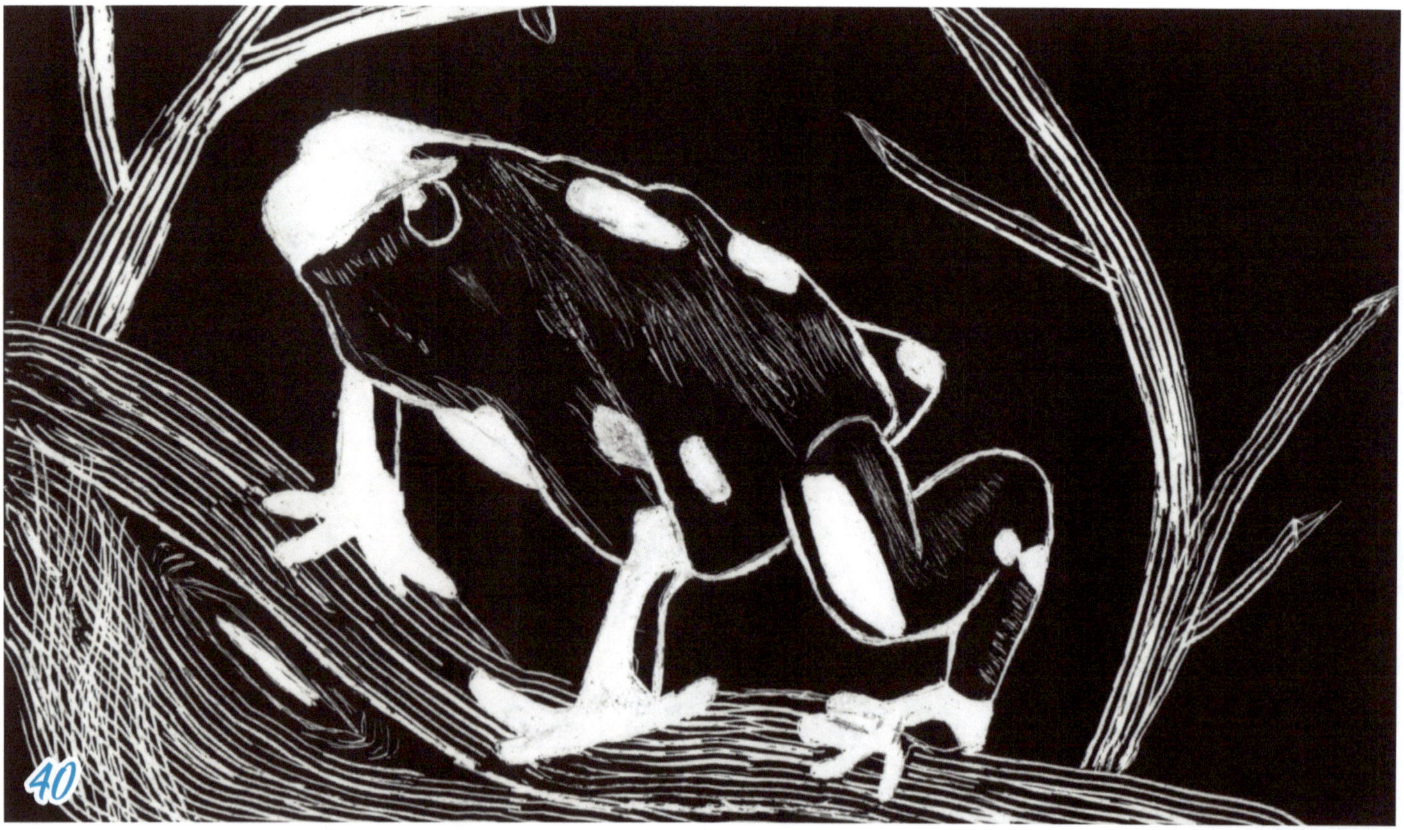

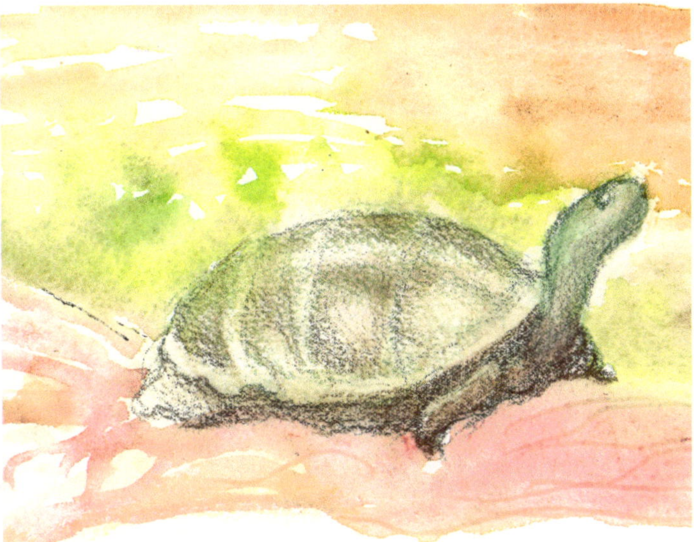
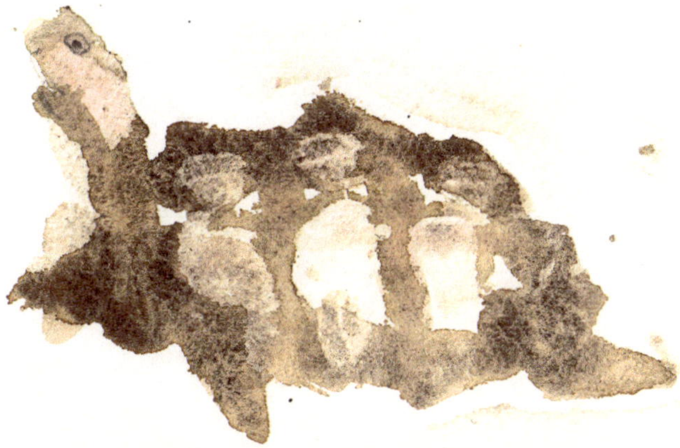
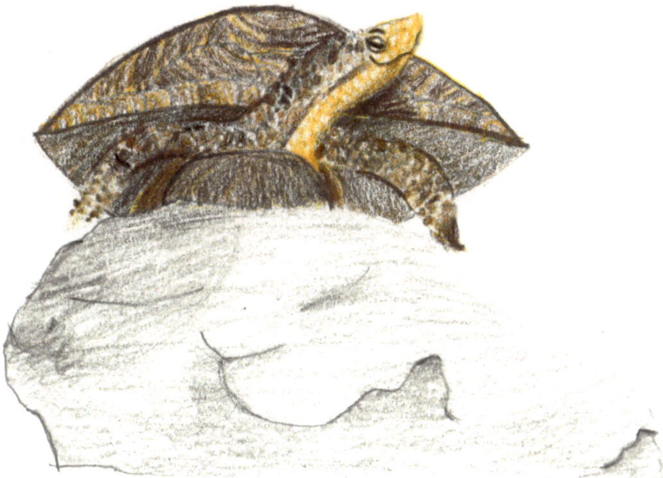
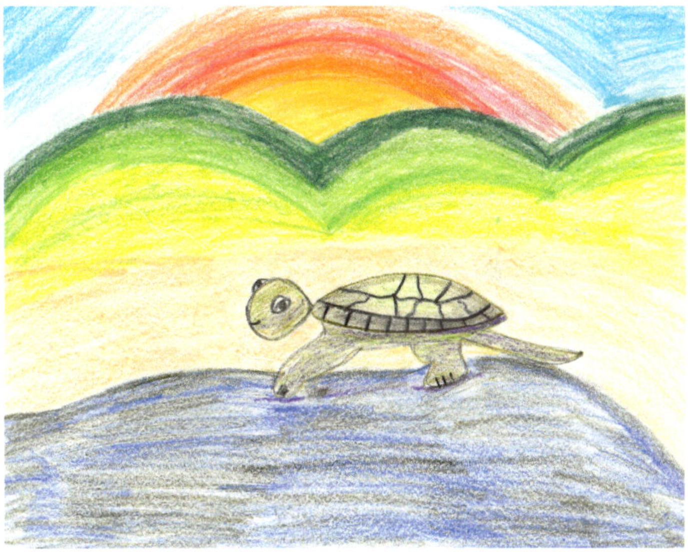

ABBIT • RACCOON • RAT • SQUIRREL • INSECTS • BEE • BLACK WI

PIDER • BUTTERFLY • COCKROACH • CRANE FLY • DEERFLY • DRAGO

FLY • MOSQUITO • MOTH • PRAYING MANTIS • FANTASY • BIR

LACKBIRD • CANADA GEESE • CHICKEN • CROW • DOVE • DUCK • E

EGRET • FINCH • FLICKERS • HAWK • HERON • HUMMINGBIRD • MA

MOCKINGBIRD • OWL • SCRUB JAY • STARLING • SWALLOW • TANA

TURKEY • TURKEY VULTURE • LANDSCAPE • CLOUDS • DELTA BREE

LATNESS • FOG • NORTH WIND • PUTAH CREEK • SUNSET • PLANTS • AC

BERMUDA GRASS • ELDERBERRY • BUCKEYE • BULBINE • CEANOTH

ENTURY PLANT • CHINESE PISTACHE • CHINESE TALLOW • COTTONW

DAFFODIL • DATURA • DISCARDED CHRISTMAS TREE • ELDERBE

FIG • FLOWERING QUINCE • FOXTAIL GRASS • GINKGO• HACKBER

EUCHERA • IRIS • JASMINE • LEAF PILE • LEMON • LUPINE • MILKWE

ISTLETOE • MODESTO ASH • MUSTARD • OLIVE TREE • ORNAMENTAL P

PENSTEMON • PERSIMMON • PINE • POMEGRANATE • POPPY • PU

EEDLE GRASS • RED HOT POKER • REDBUD • REDWOOD • ROSEMA

ALVIA • SOURGRASS • SUNFLOWER • SYCAMORE • TOMATO • TOYON •

VALLEY OAK • WALNUT • FISH AND REPTILES • FROG • SALMON • TU

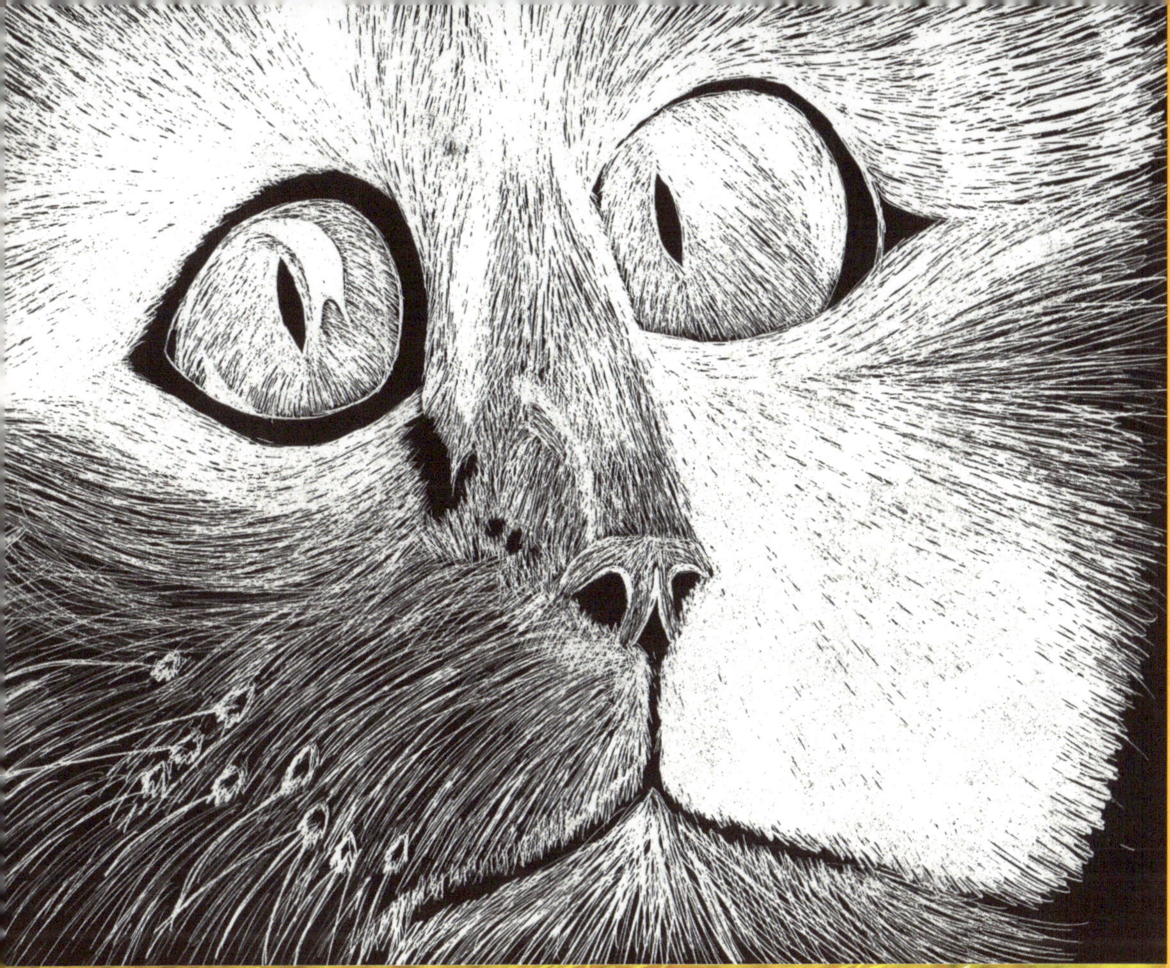

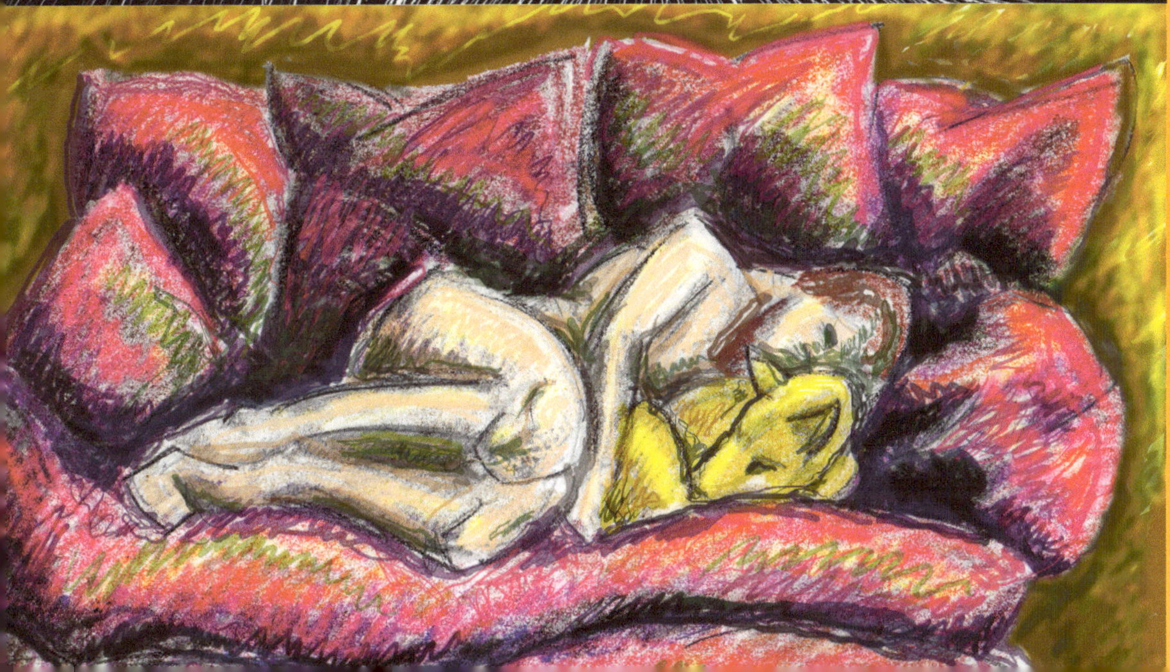

MAMMALS

The following artists created the artwork in this section. The list is alphabetical by category and does not necessarily represent the order nor page number the artwork appears; please refer to the page numbers next to the category to locate where the art appears. Some teachers/schools did a class project (CP) for one category; all the artists involved in that project are listed in italics under the teacher and school name within the category. For other class projects the artist chose their category; their names appear under that category with a footnote indicating their teacher/school. If there is a number after an artist's name they provided more than one piece of art for that category.

BAT
(pages 44)
Porterfield, Jessie

CAT
(pages 43 and 50)
Burns, Anna Lucia - CP[1]
Daniels, Emma - CP[1]
Edwards, Alexa - CP[1]

COW
(page 46)
Brown, Marie-Therese
Frisch, Gabriella - CP[1]
Hayes, Lillian - CP[1]

COYOTE
(page 47)
Benson, James - CP[2]
Dhillon, Amrita - CP[1]
Kiyama, Amy - CP[1]
Panson, Sarah - CP[1]

DOG
(pages 43 and 48)
Green, Pam (2)
Ha, Maya - CP[1]
Munn, Abby - CP[1]

HORSE
(page 49)
Flanders, Marian

OTTER
(pages 48-49)
Geary , Truth
Grehm, Hannah - CP[1]
Raminez, Brittany - CP[1]
Simi, Alden - CP[2]

OPOSSUM
(page 45)
Flanders, Marian (2)

RABBIT
(pages 50-51)
Clark , Bailey - CP[1]
Friesen, Julie
Hayes, Lillian - CP[1]
Madison, Jamie
Moon, Bailey - CP[1]
Schaedler, Ava - CP[1]

RACCOON
(pages 52-53)
Ayotte, Erin
Nelson, Natalie
CP, 3rd, Harriet
Gadisman, Patwin
Elementary
Amilia
Ben
Calla
Casey
David
Deema
Elias
Emma
Estella
Esther
Haley
Harriet
Isabela
Jasarae
Jonah
Jonas
Kelly
Nico
Tobin
Wil
Wren

RAT
(page 45)
Hajian, Soheila
Mahnoor, Khan - CP[1]
Riley, Abigail - CP[1]
Smith, Hannah - CP[1]

SQUIRREL
(page 50)
Love, Christine - CP[1]
Schraner, Brandt
Wagnon, Sydney - CP[1]

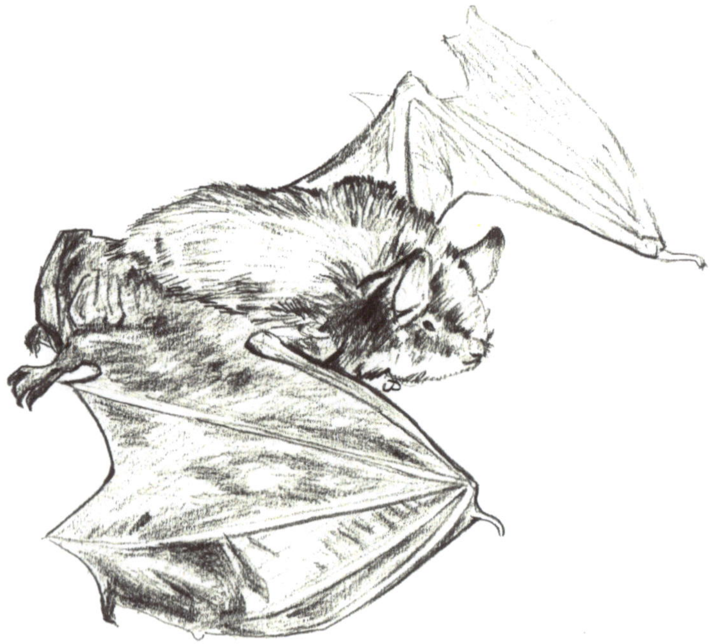

[1]Student's Choice Class Project, Teacher: Melanie Bown, St. Francis High School • [2]Student's Choice Class Project, Teacher: Carrie Ceniseroz, St. James School

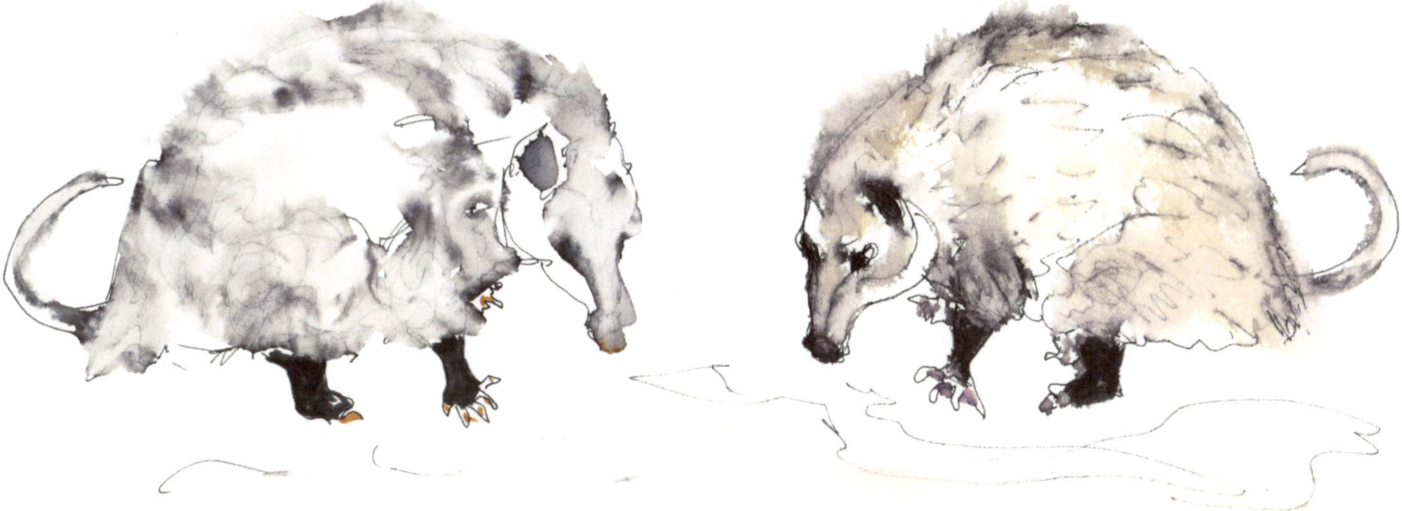

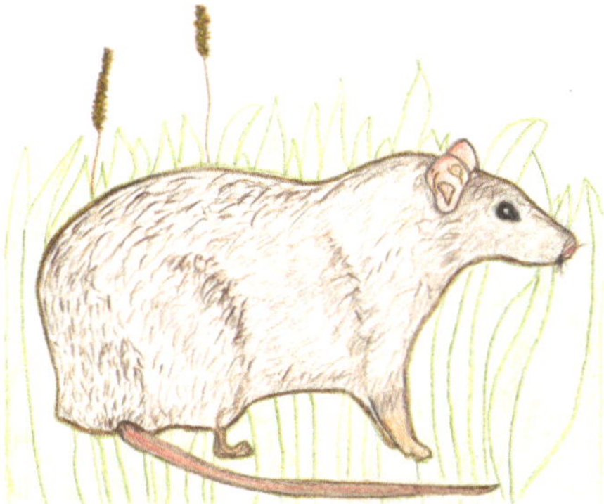

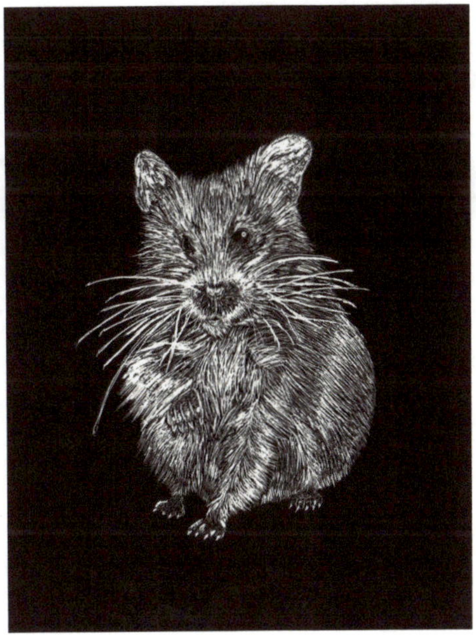

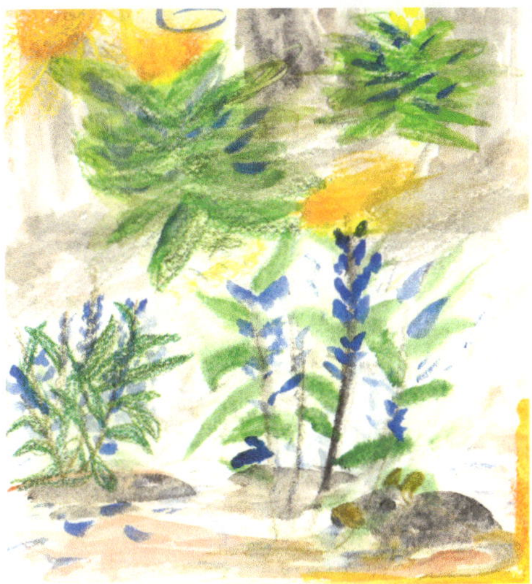

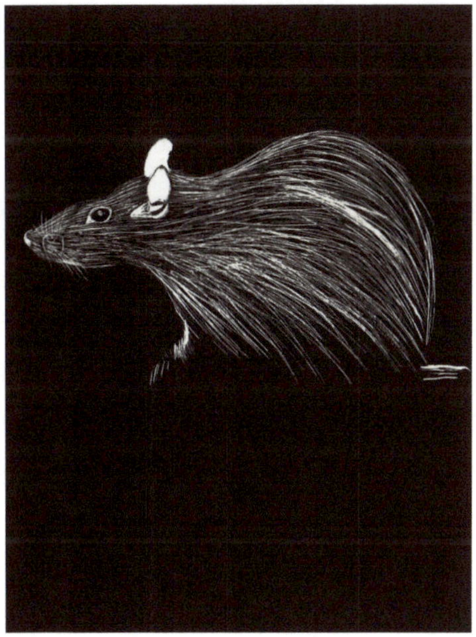

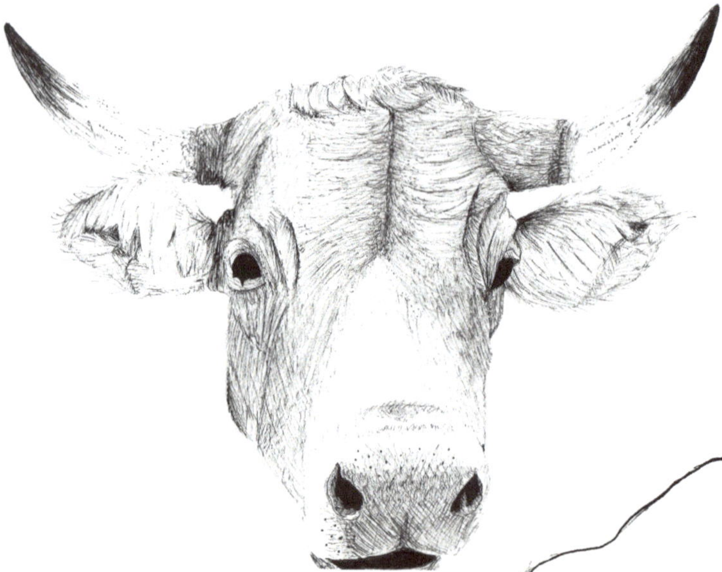

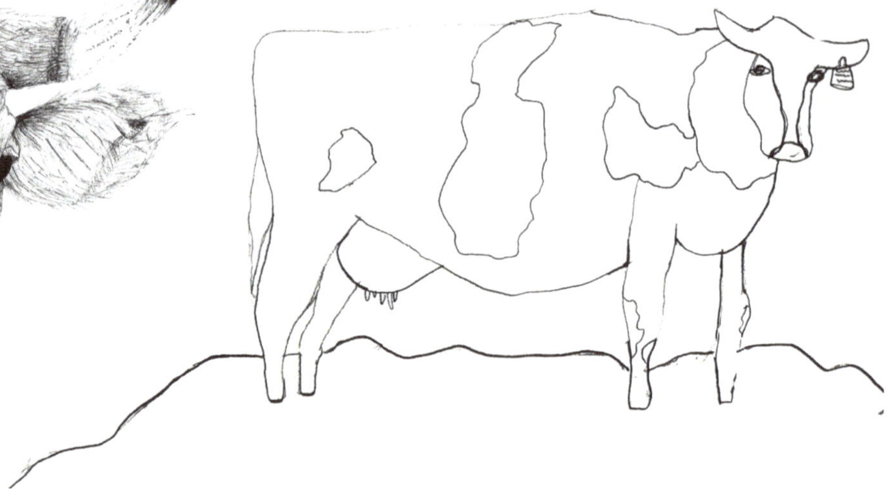

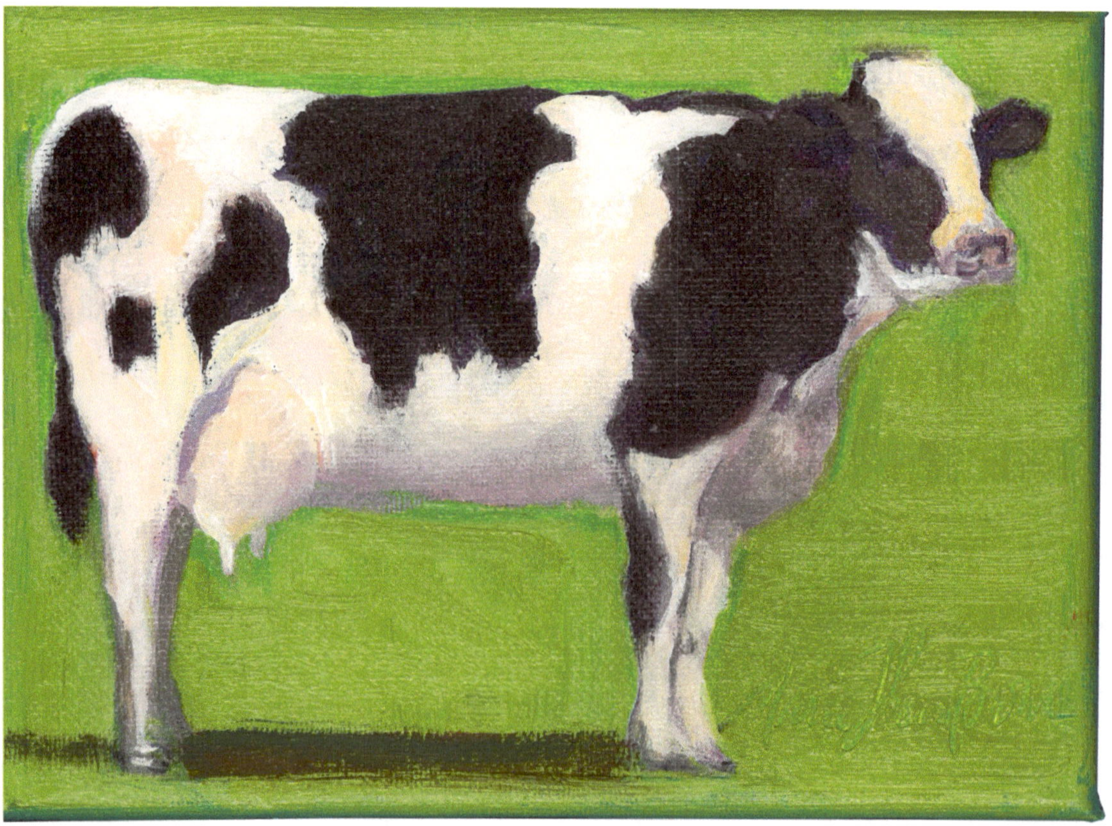

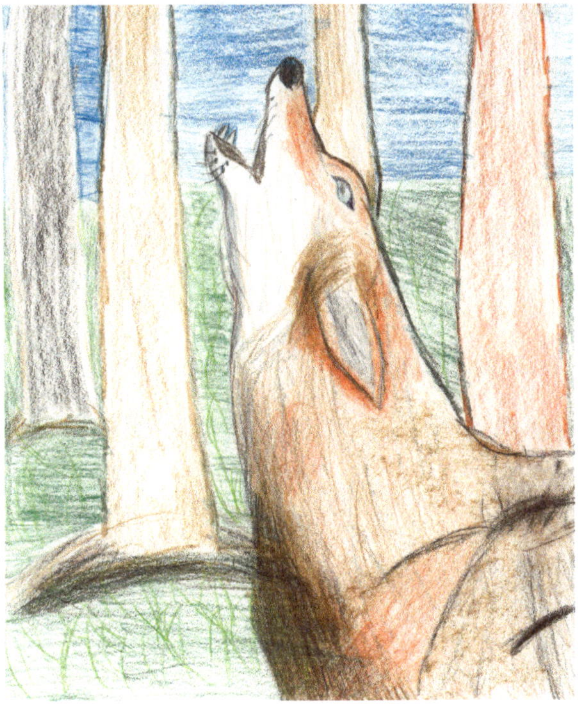
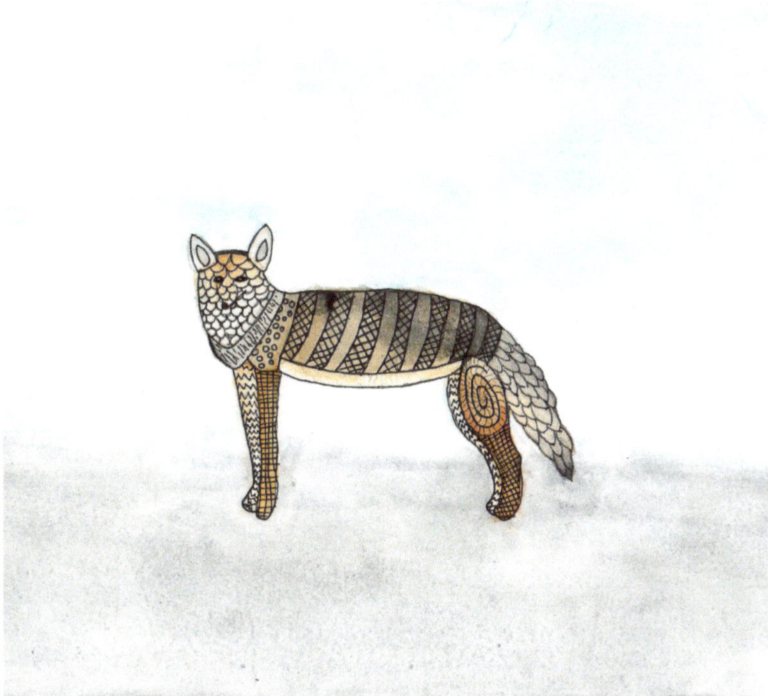
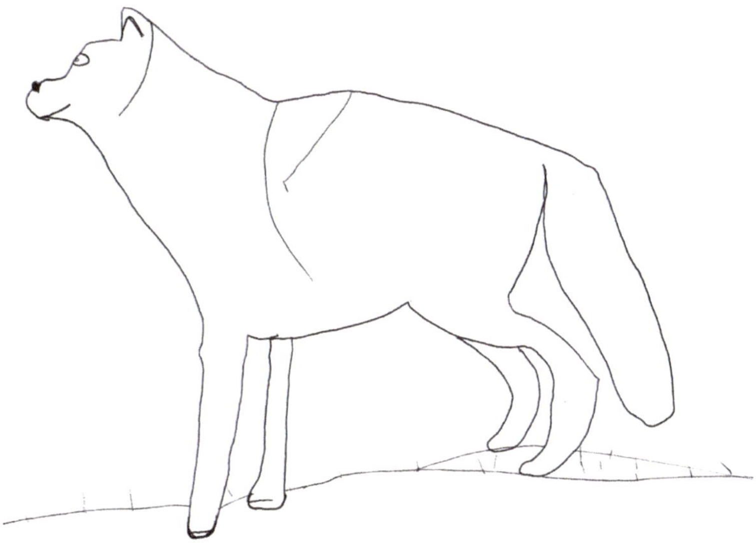
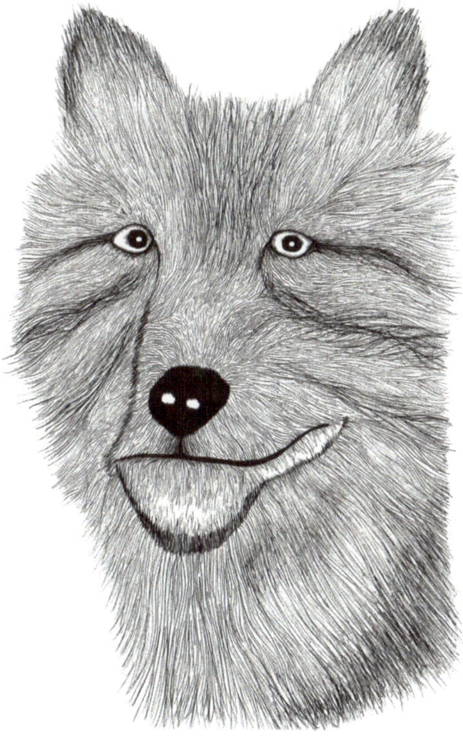

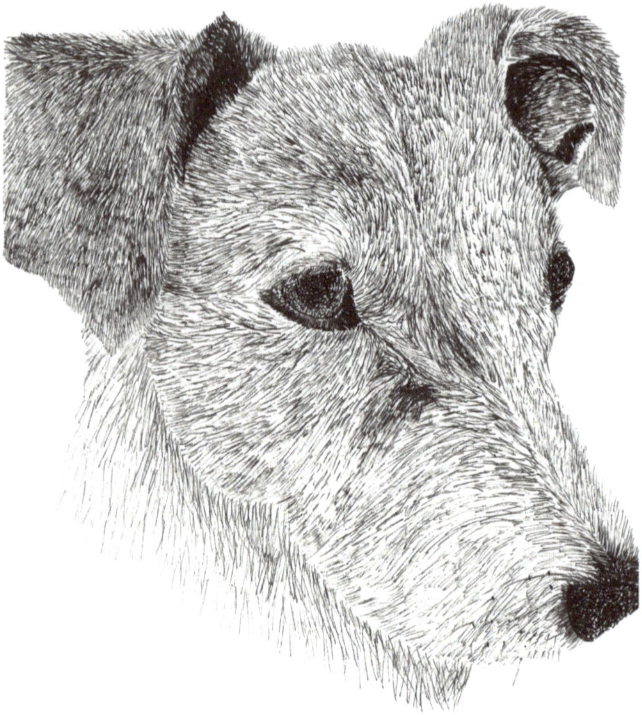

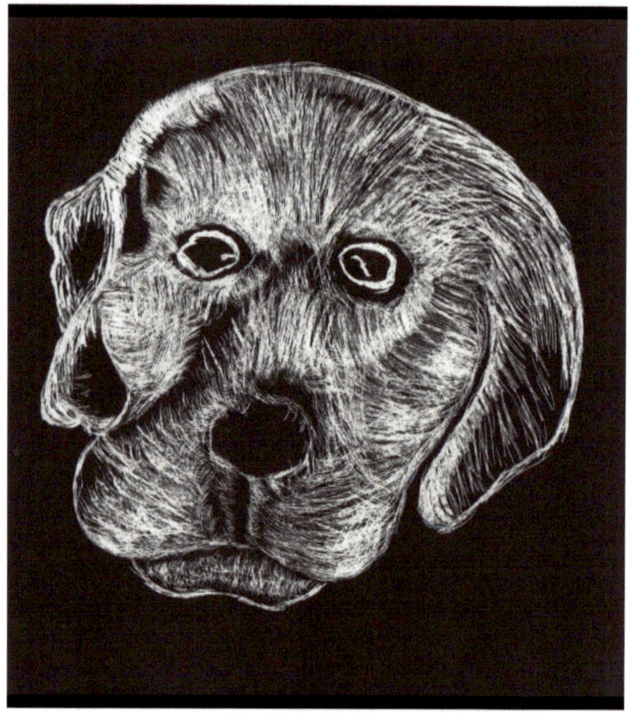

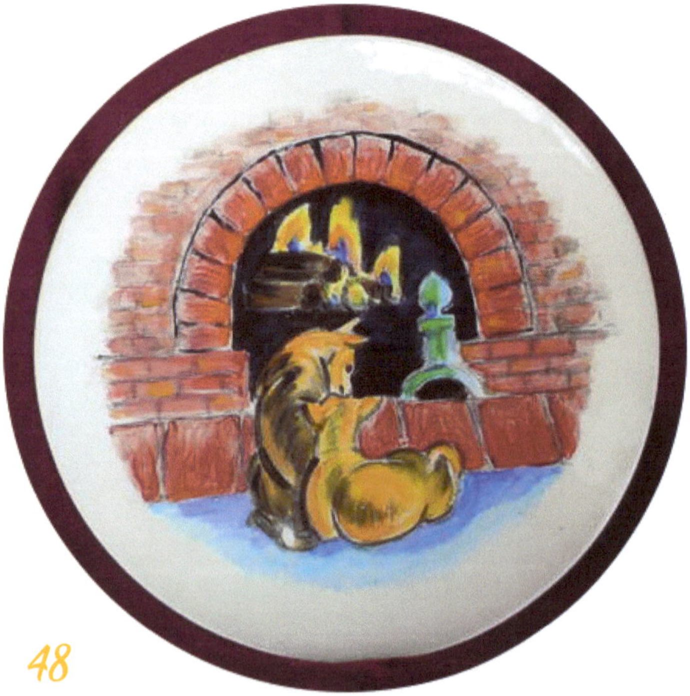

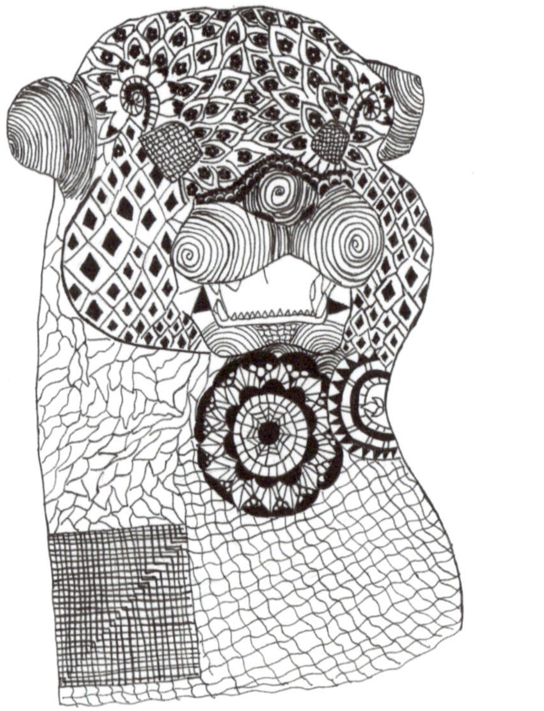

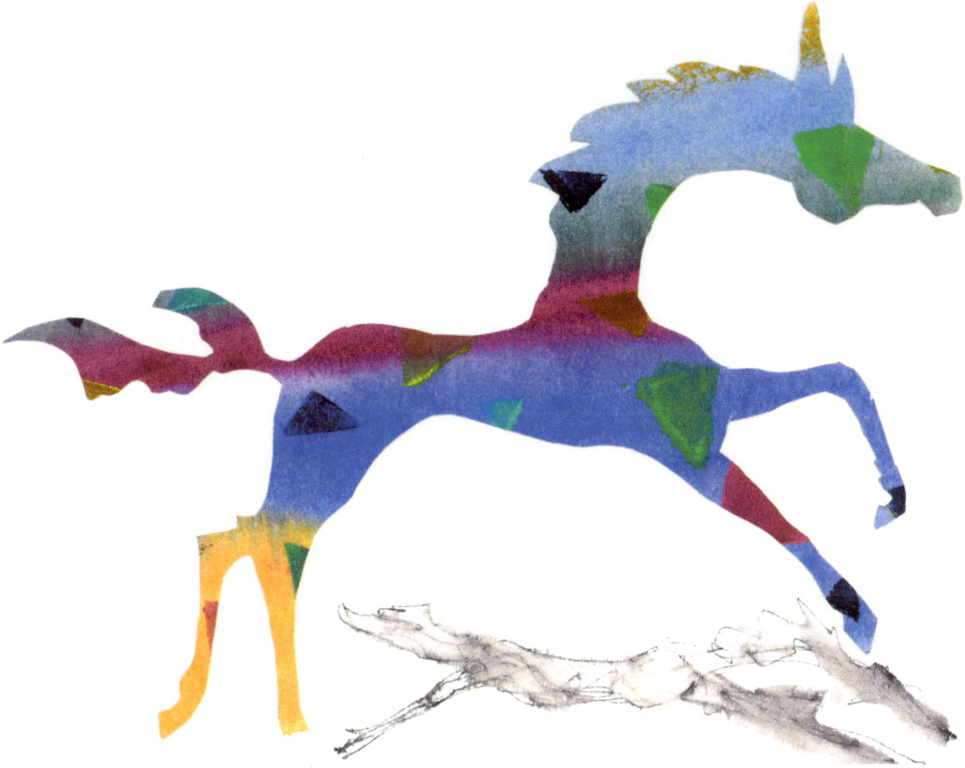

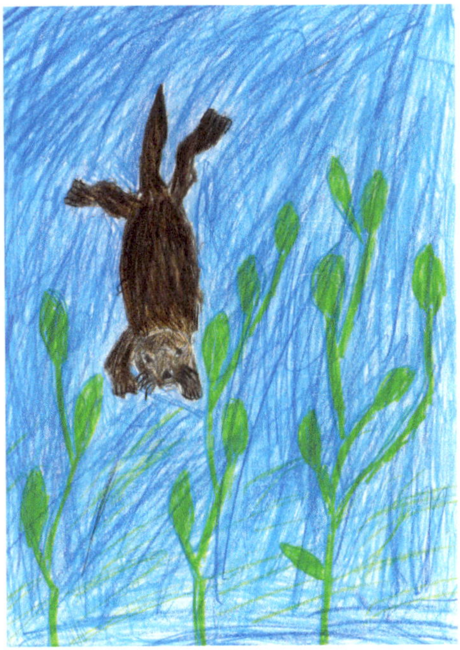

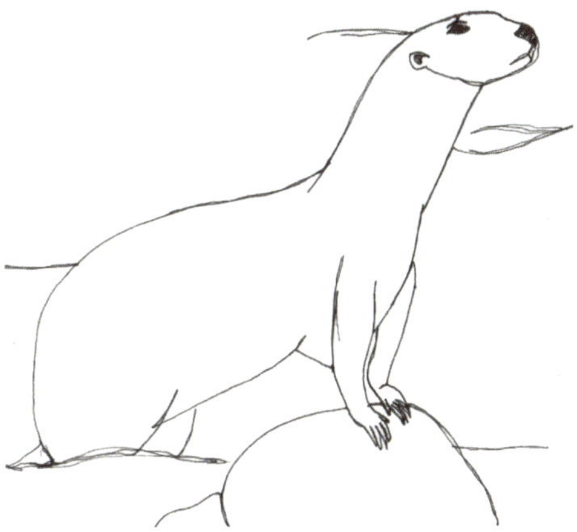

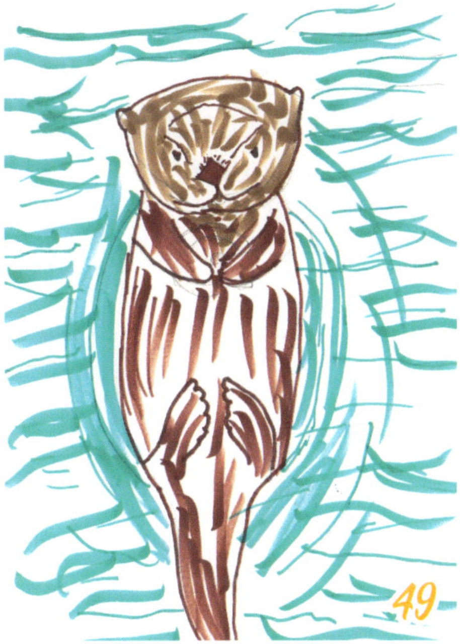

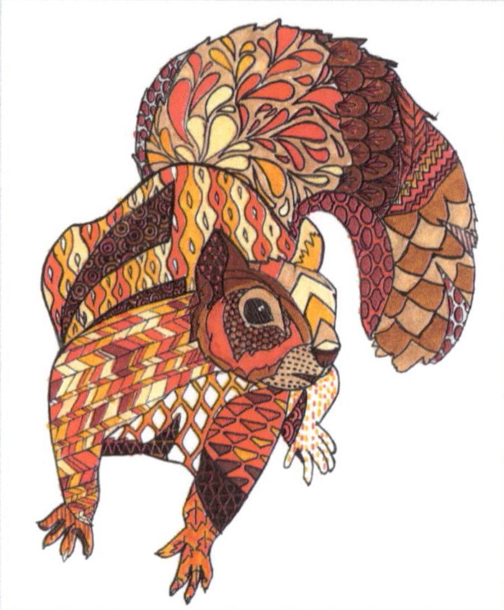
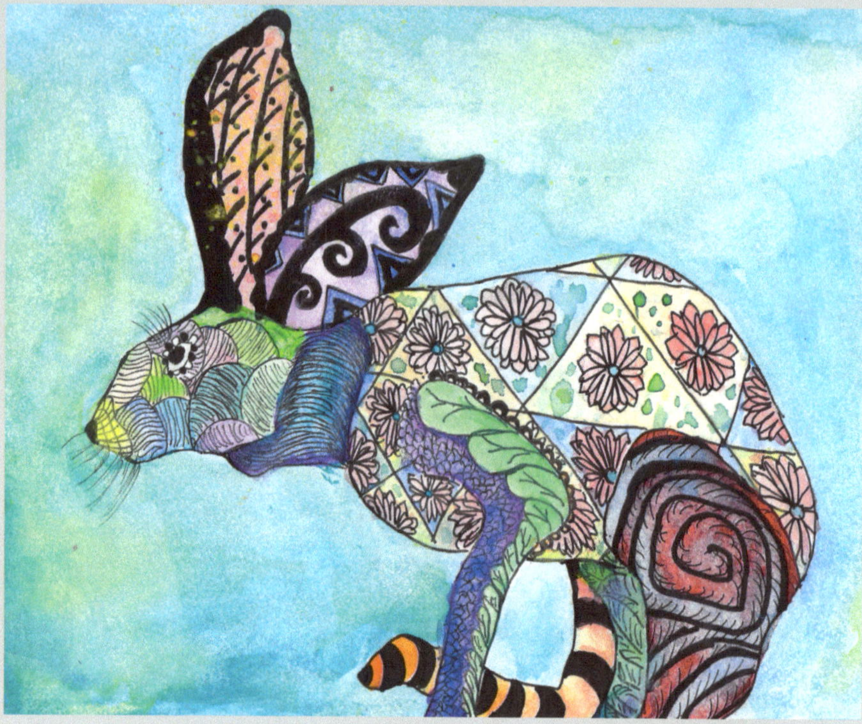
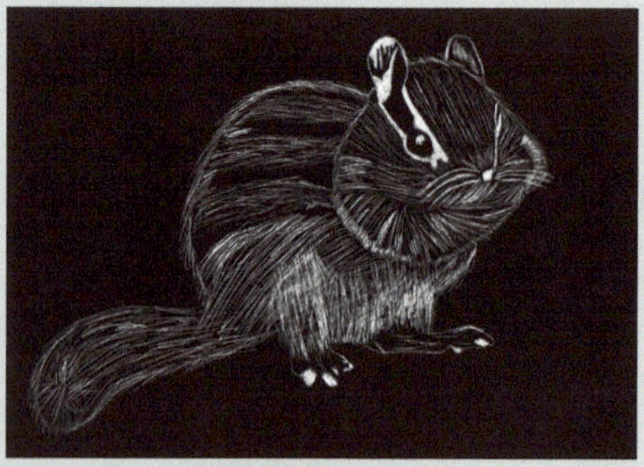
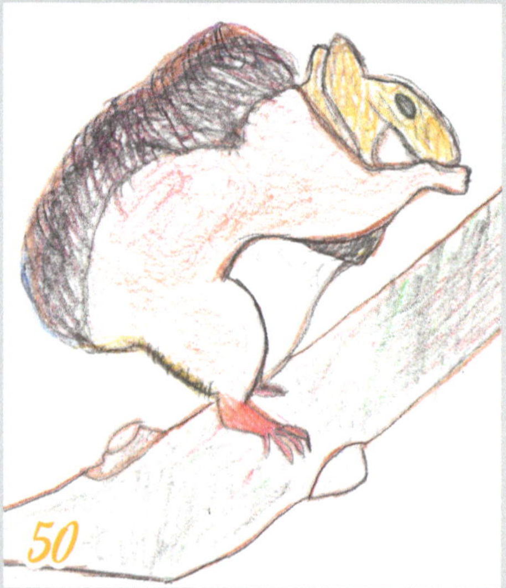
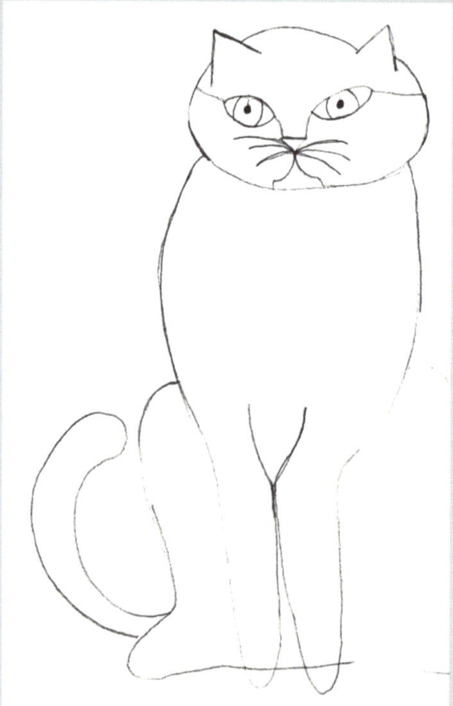
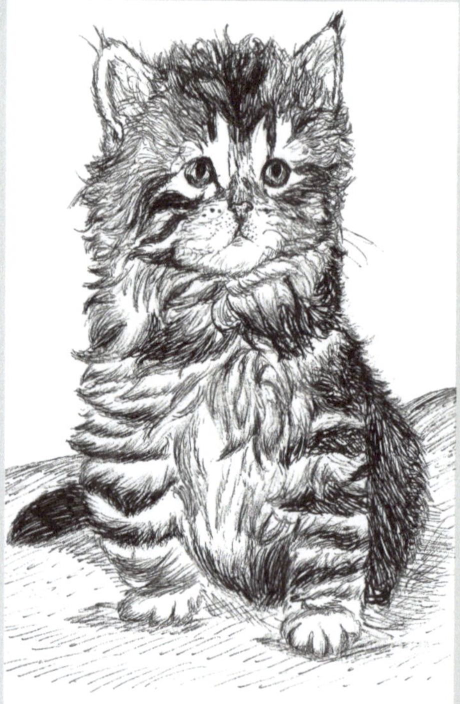

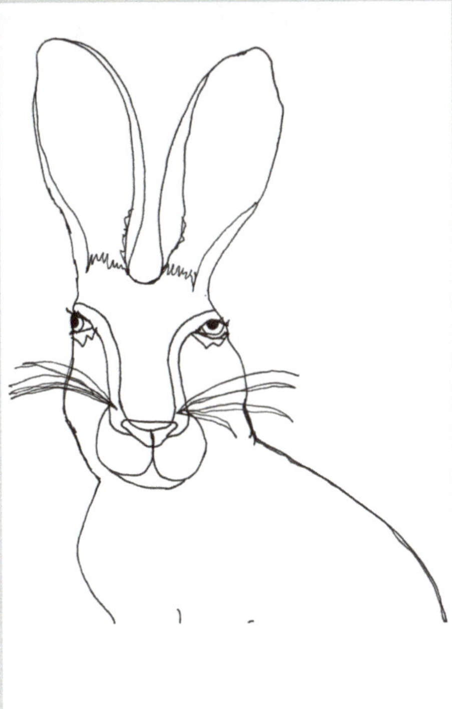
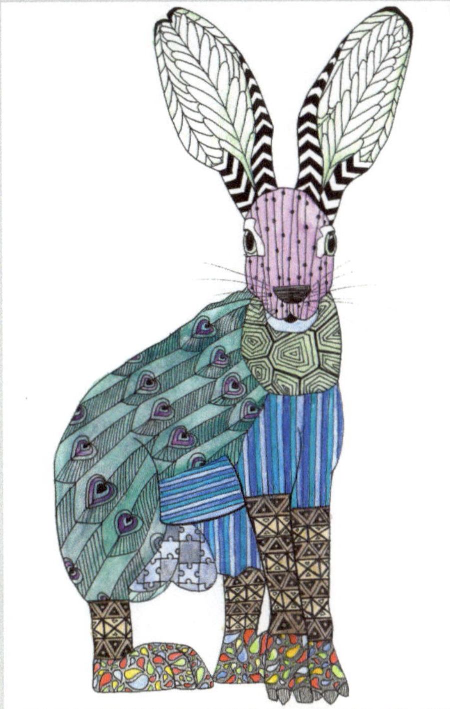
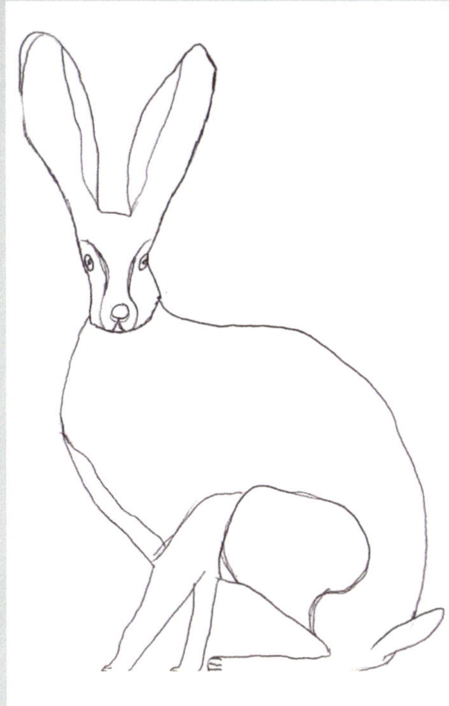
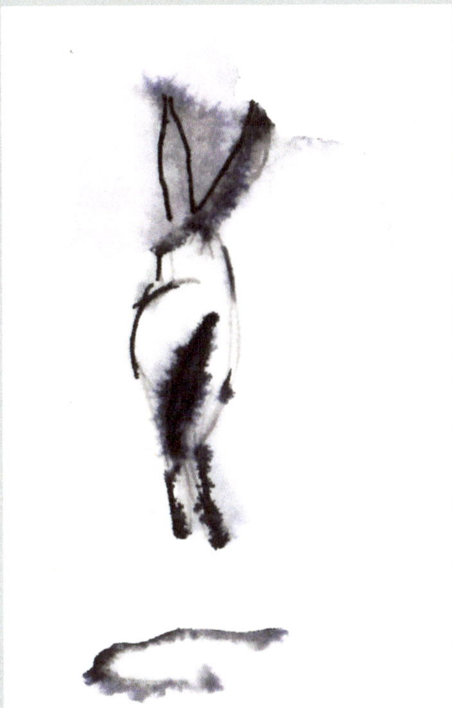
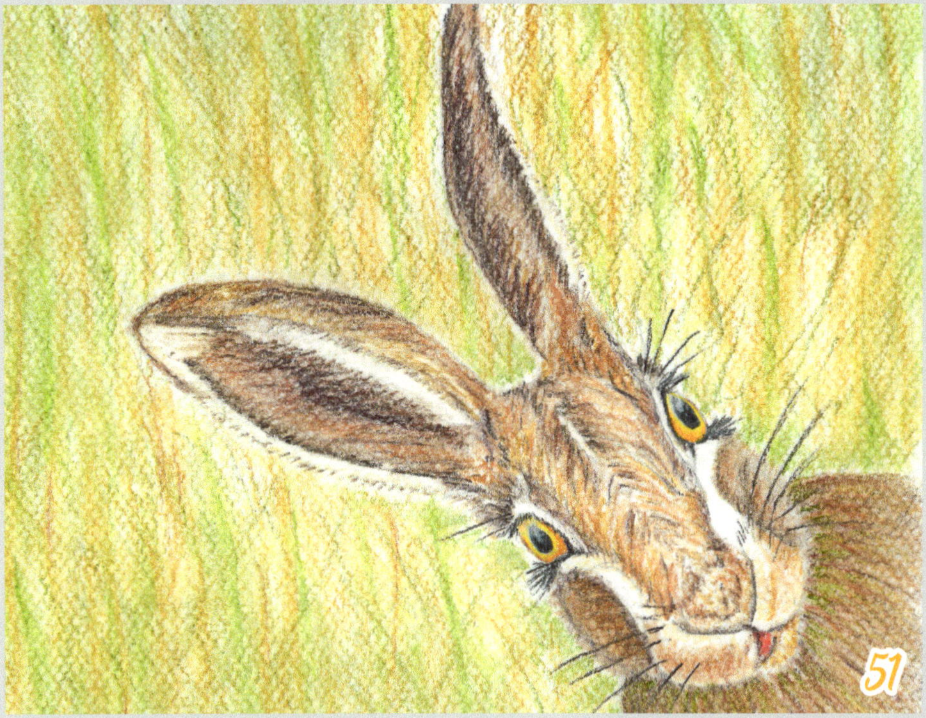

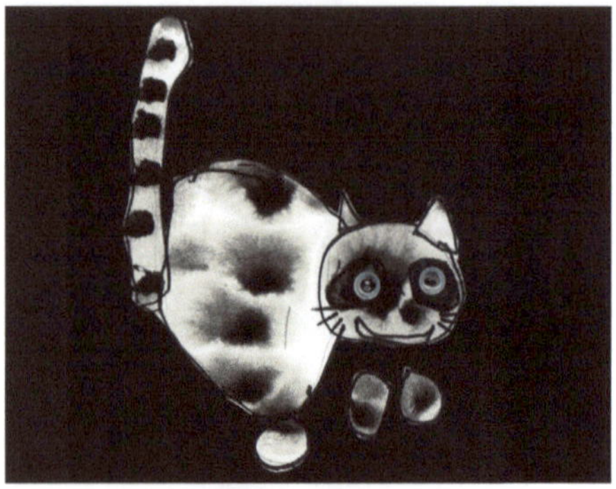
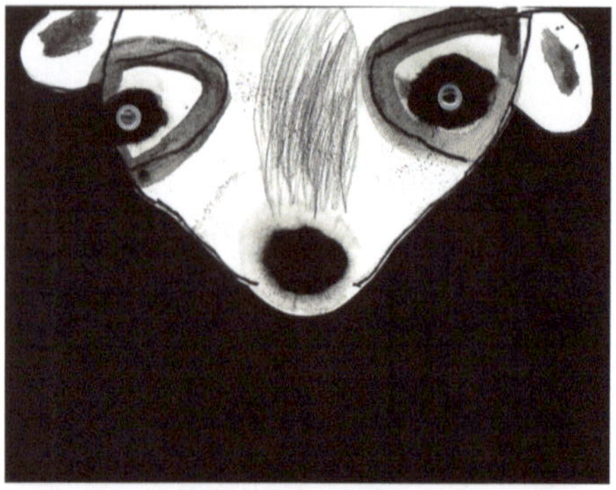
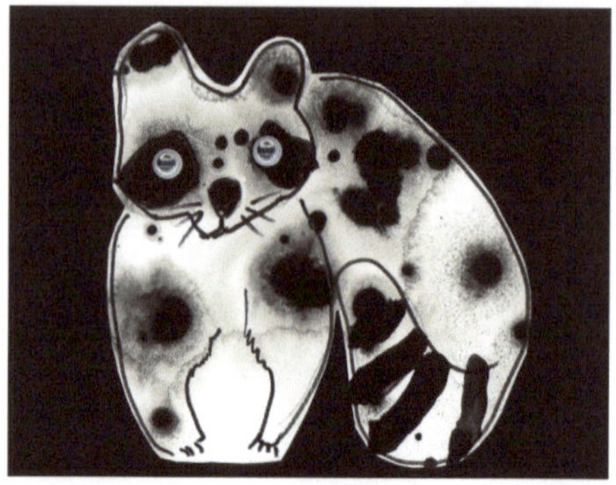
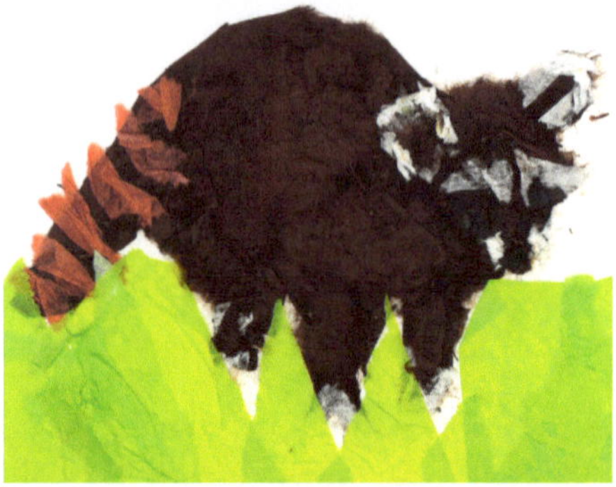
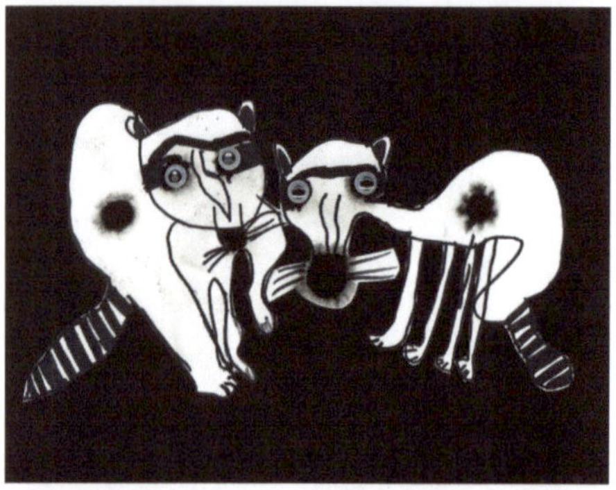
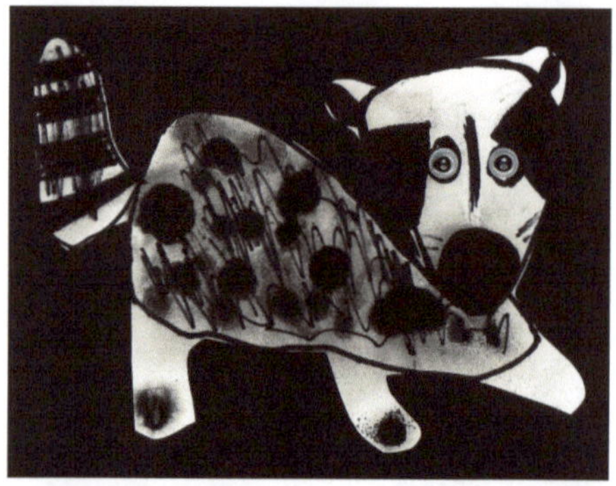
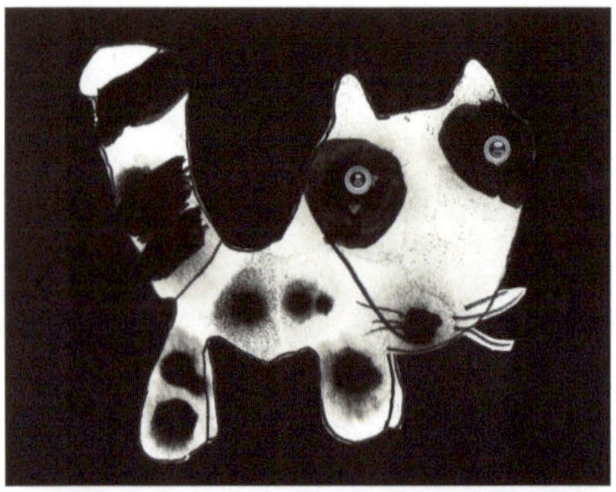
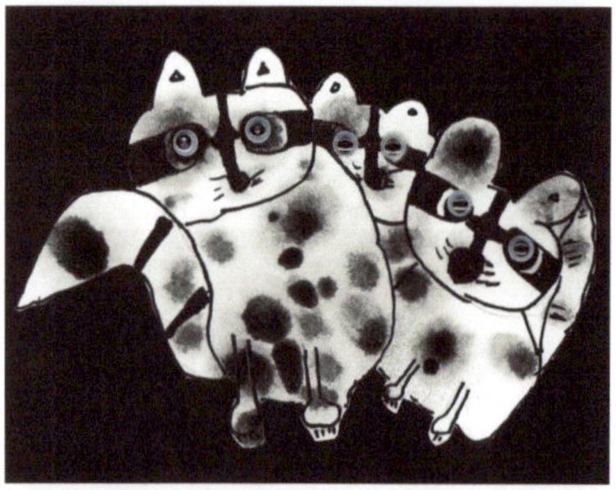
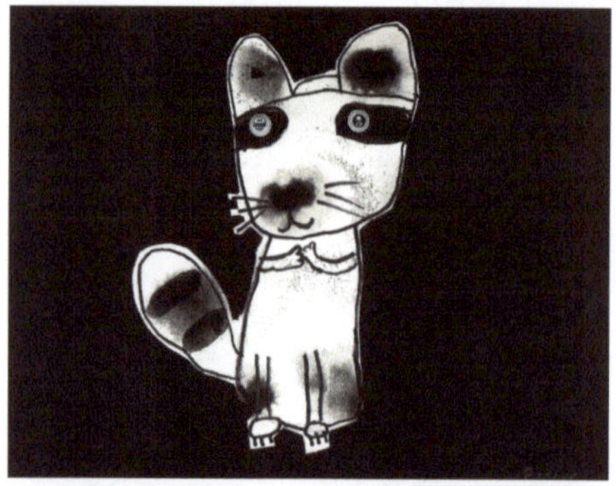
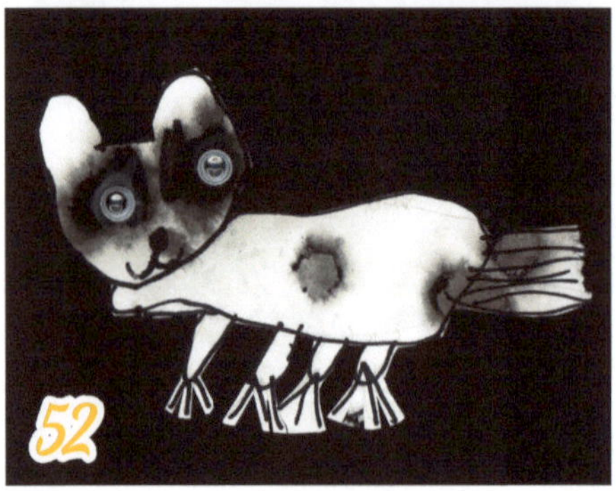
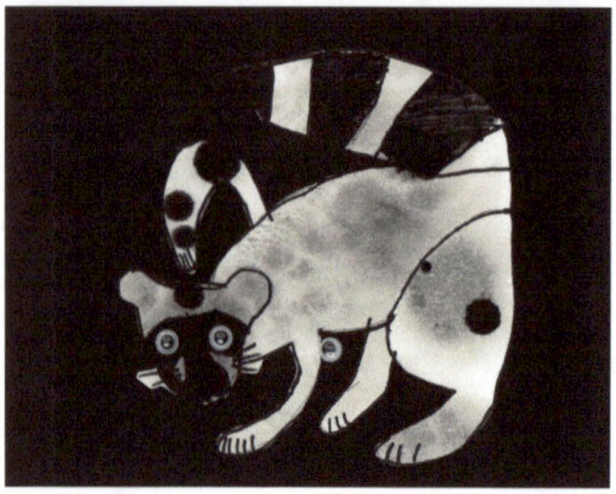
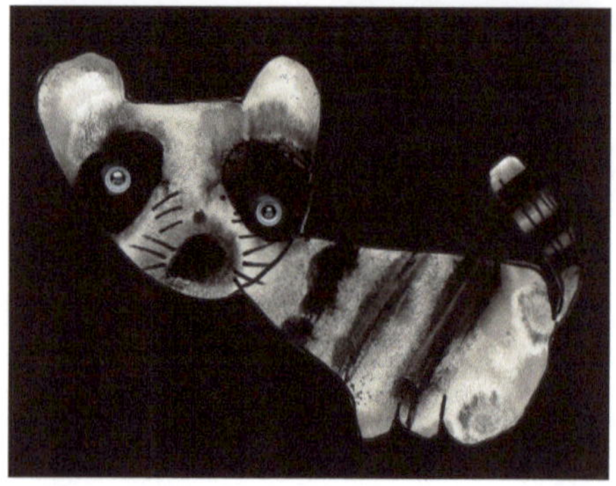

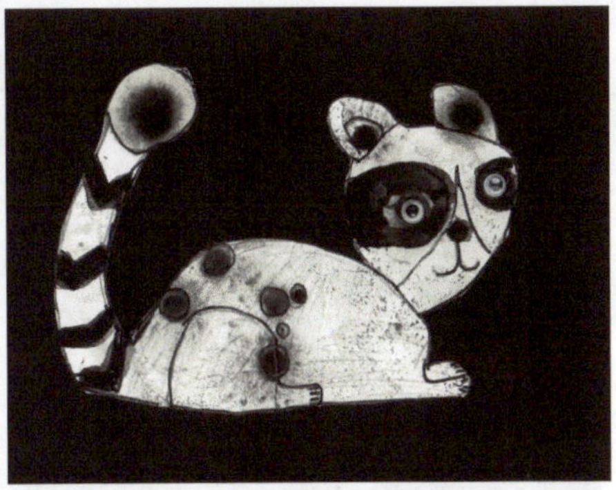
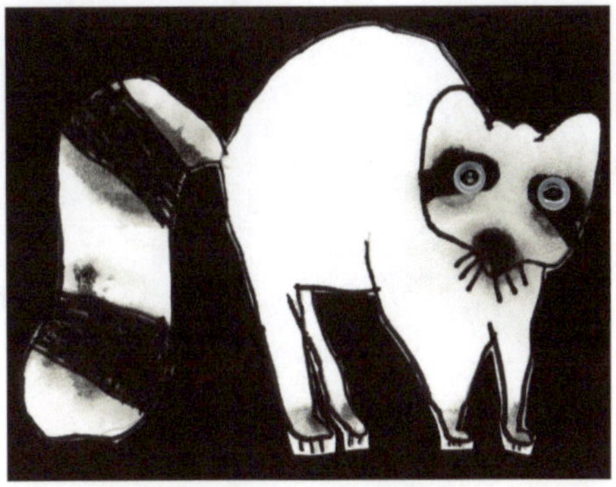
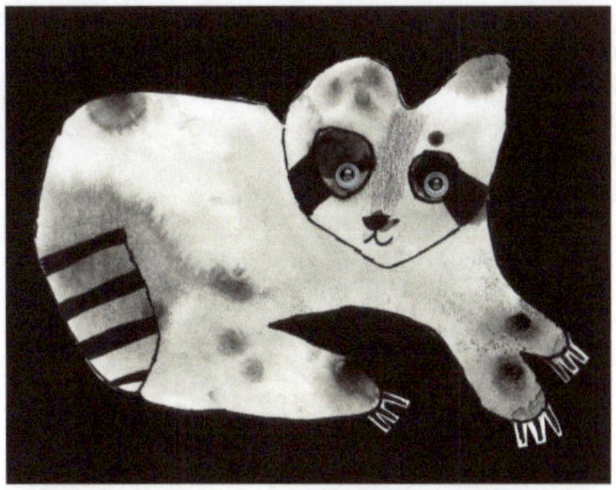
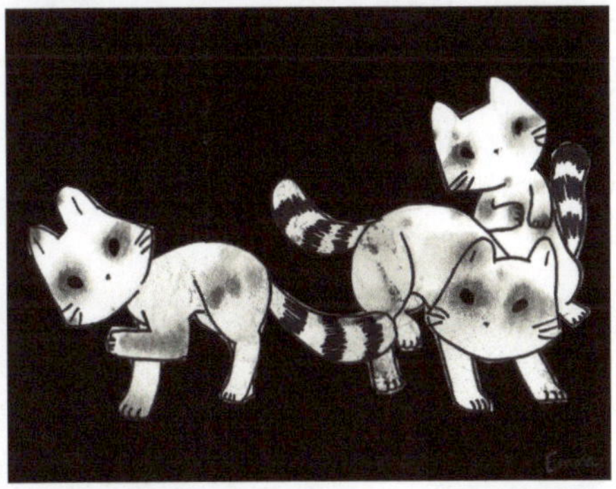
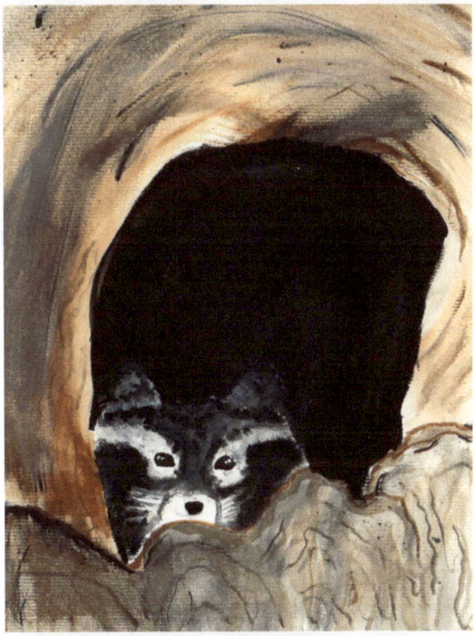
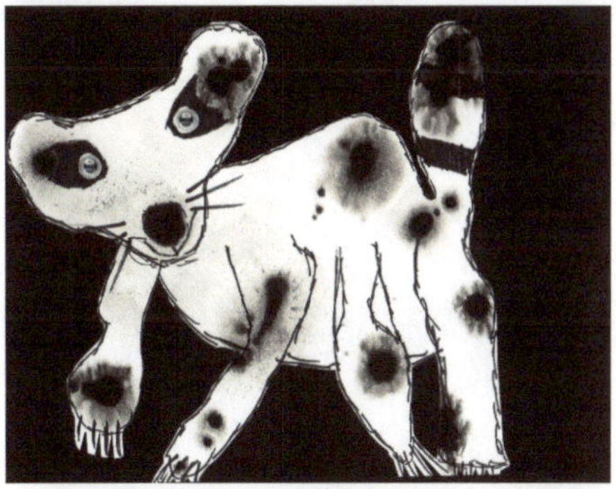
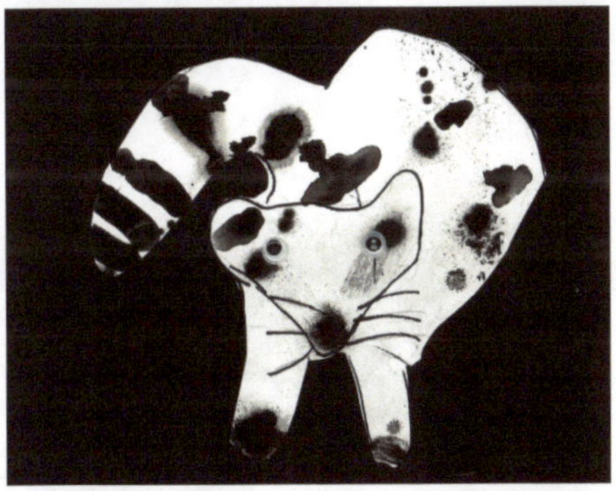
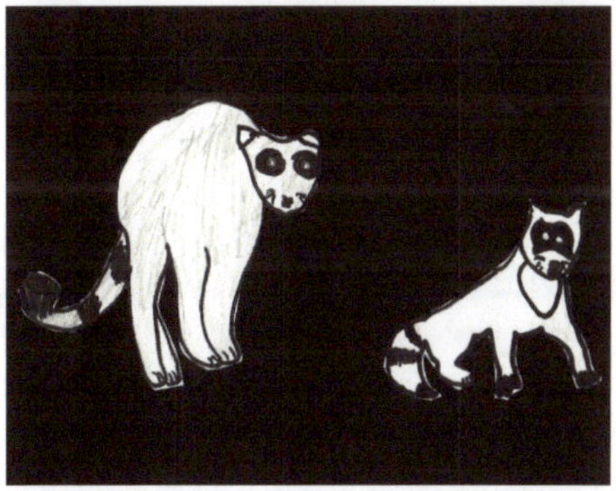
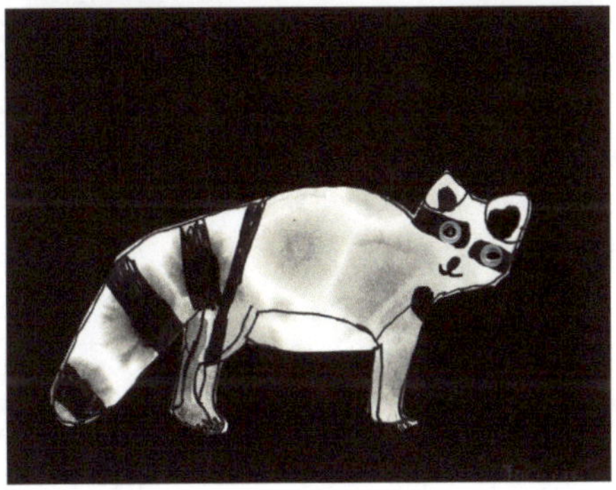
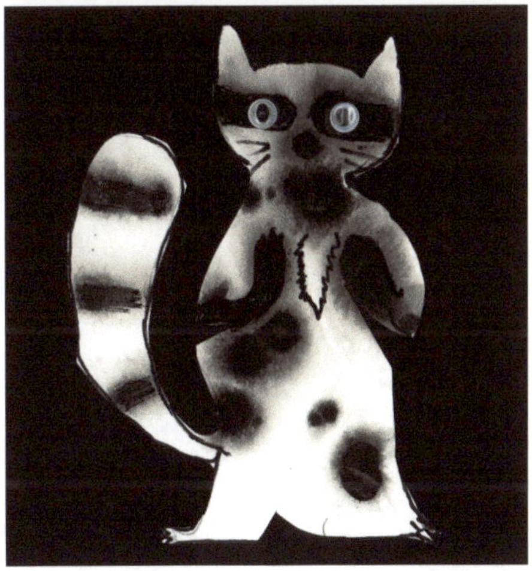
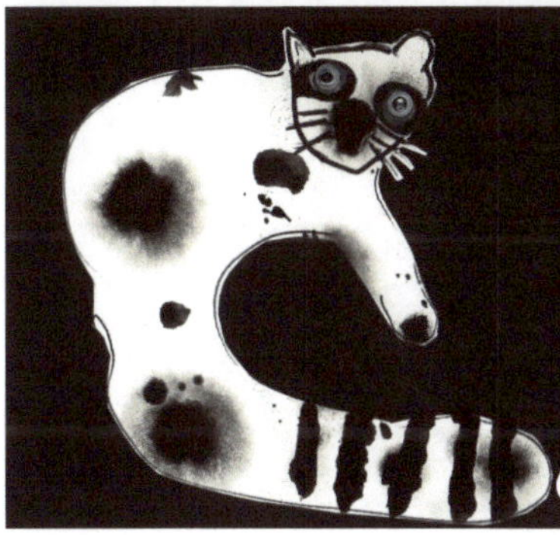

NSECTS · BEE · BLACK WIDOW SPIDER · BUTTERFLY · COCKROACH · CR

LY · DEERFLY · DRAGONFLY · FLY · MOSQUITO · MOTH · PRAYING MA

FANTASY · BIRDS · BLACKBIRD · CANADA GEESE · CHICKEN · CRO

OVE · DUCK · EAGLE · EGRET · FINCH · FLICKERS · HAWK · HER

UMMINGBIRD · MAGPIE · MOCKINGBIRD · OWL · SCRUB JAY · STAR

SWALLOW · TANAGER · TURKEY · TURKEY VULTURE · LANDSCA

LOUDS · DELTA BREEZE · FLATNESS · FOG · NORTH WIND · PU

REEK · SUNSET · PLANTS · ACORN · BERMUDA GRASS · ELDERBER

UCKEYE · BULBINE · CEANOTHUS · CENTURY PLANT · CHINESE PISTA

CHINESE TALLOW · COTTONWOOD · DAFFODIL · DATURA · DISCAR

HRISTMAS TREE · ELDERBERRY · FIG · FLOWERING QUINCE · FOX

RASS · GINKGO· HACKBERRY · HEUCHERA · IRIS · JASMINE · LEAF

LEMON · LUPINE · MILKWEED · MISTLETOE · MODESTO ASH · MUST

OLIVE TREE · ORNAMENTAL PEAR · PENSTEMON · PERSIMMON ·

POMEGRANATE · POPPY · PURPLE NEEDLE GRASS · RED HOT POK

EDBUD · REDWOOD · ROSEMARY · SALVIA · SOURGRASS · SUNFLOW

YCAMORE · TOMATO · TOYON · TULE · VALLEY OAK · WALNUT · FISH

EPTILES · FROG · SALMON · TURTLE · MAMMALS · CAT · COW · COYO

OG · HORSE · OTTER · OPOSSUM · RABBIT · RACCOON · RAT · SQUIR

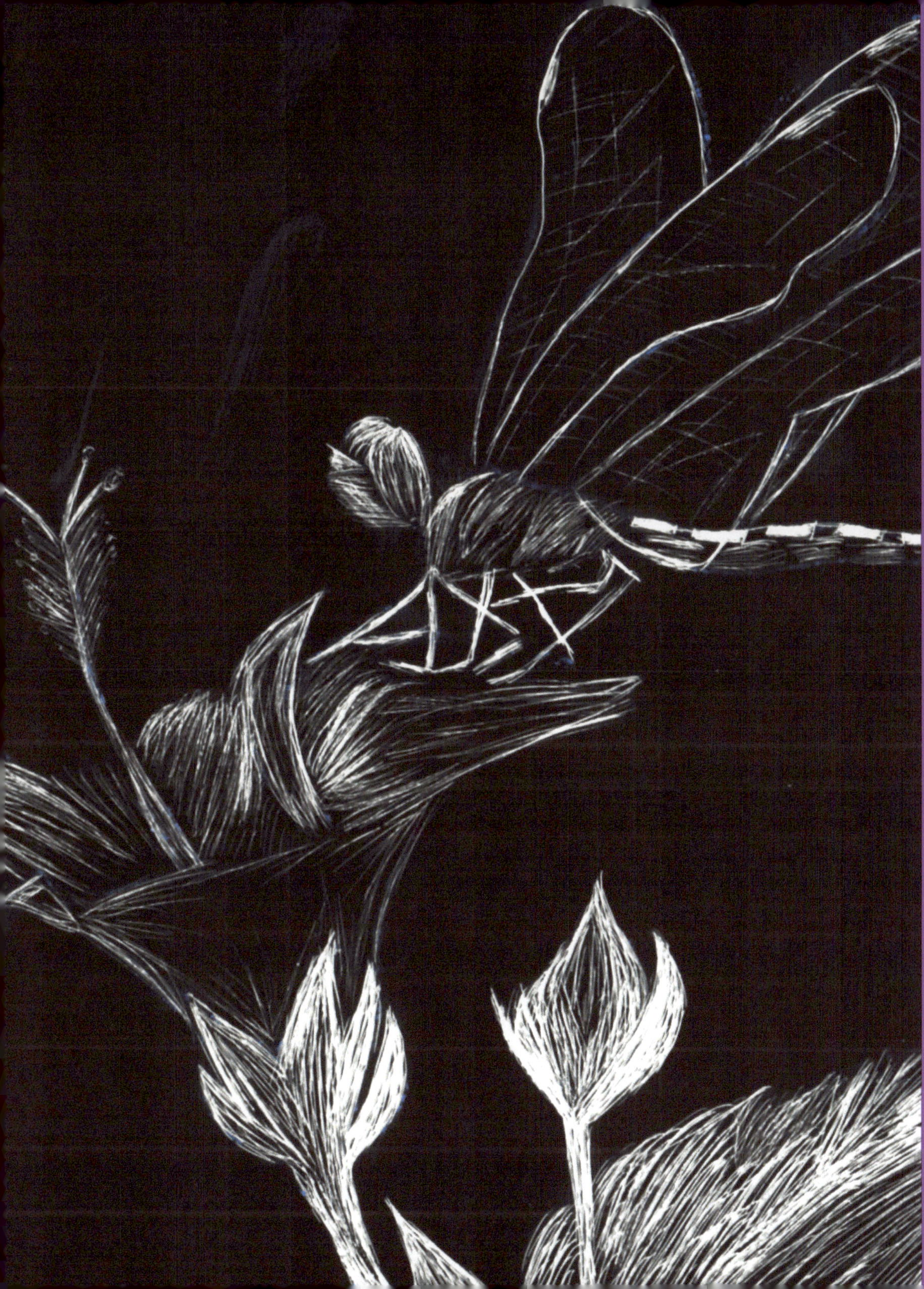

INSECTS

The following artists created the artwork in this section. The list is alphabetical by category and does not necessarily represent the order nor page number the artwork appears; please refer to the page numbers next to the category to locate where the art appears. Some teachers/schools did a class project (CP) for one category; all the artists involved in that project are listed in italics under the teacher and school name within the category. For other class projects the artist chose their category; their names appear under that category with a footnote indicating their teacher/school. If there is a number after an artist's name they provided more than one piece of art for that category.

BEE
(page 57)
Briggs, Taylor - CP[1]
Duran, Christian - CP[2]
Espinosa, Matteo - CP[2]
Momsen, Skylar - CP[1]
Valdez, Avonlea - CP[1]

BLACK WIDOW SPIDER
(page 66)
Malek, Hannah - CP[1]

BUTTERFLY
(pages 58-65)
Frerichs, Stacie (2)
Kirk, Nancy
Maze, Stella
Vetter, Stacey
Wilhelm, Barbara
CP, 1st, Carrie Ceniseroz, St. James Elementary
Aldrete, Chloe
Asmundsen, Zach
Chiasson-Larson, Damien
Delaini, Domenico
Dennis, Matthias
Dewar, Colton
Dott, Amy
Falon, Peter
Ferrick, Luke
Gorospe, David
Graham, Kristian
Grammer, Henry
Hong, Olivia
Kelly, John
Kreylos, Athena
Landry, Jonah
Lynch, Rowan
Mariani, Angelina
Mariani, Grace
Mayhew, Reece
McGan, Delaney
Mendez-Nunez, Ana
Montoya, Sophia
Murray, Kevin
Myers, Kate Ann
Park, Jane
Ramos, Thomas
Sale, Saveria
Thompson, Christopher
Vallero, Matthew
Watkins, Ashlyn
Williams, Austin
CP, Melanie Bown, St. Francis High School
Barger, Grace
Campos, Andrea
Cordano, Kristi
Dahdouh, Julionah
Davis, Brielle
Deak, Kiera
Dorsey, Hannah
Guerrero, Karla
Hayes, Lillian
Konigsfeld, Abby
Leon, Karissa
Mudd, Sarah
Overholtzer, Megan
Platanesi, Lindsey
Qumar, Iza
Simpson, Jordan
CP, 4th/5th Amelia Hansen/Jennifer Nachmanoff, Birch Lane
Abby
Ai
Andres
Austin
Becket
Bella
Dina
Django
Edison
Emma
Isis
Joseph
Kaelen
Lele
Lydell
Maddie
Mariam
Max
Maya
Natalie
Owen
S. Owen
Rahan
Riley
Rylan
Sarah
Sienna
Tessas

COCKROACH
(page 67)
Nelson, Natalie

CRANE FLY
(page 66)
Wagner, Jerri

DEERFLY
(page 66)
Pauley, Cynthia

DRAGONFLY
(pages 55, 56 and 68)
Maze, Stella
Wagner, Jerri
Wilhelm, Barbara
CP, Melanie Bown, St. Francis High School
Christopulos, Marisa
Edwards, Alexa
Fargen, Emily
Grehm, Hannah
Kiyama, Amy
Overholtzer, Megan
Parish, Gabi

FLY
(page 66)
Pauley, Cynthia

MOSQUITO
(page 66)
Hebert, Nicole

MOTH
(page 69)
CP, Melanie Bown, St. Francis High School
Briggs, Taylor
Laing, Devon
Malek, Hannah
Mueller, Ellie

PRAYING MANTIS
(page 67)
BonDurant, Gayle

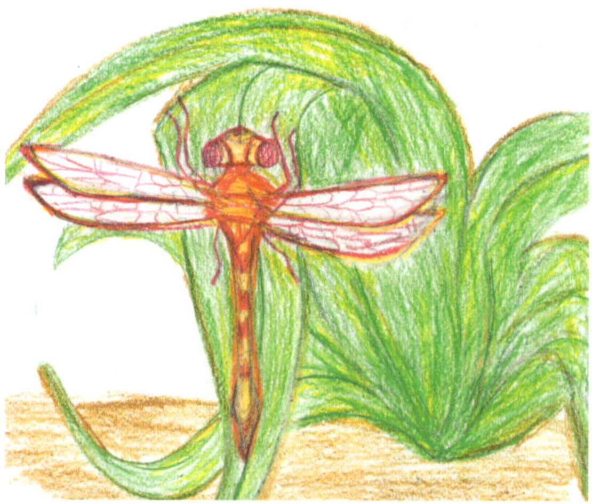

[1]Student's Choice Class Project, Teacher: Melanie Bown, St. Francis High School • [2]Student's Choice Class Project, Teacher: Carrie Ceniseroz, St. James School

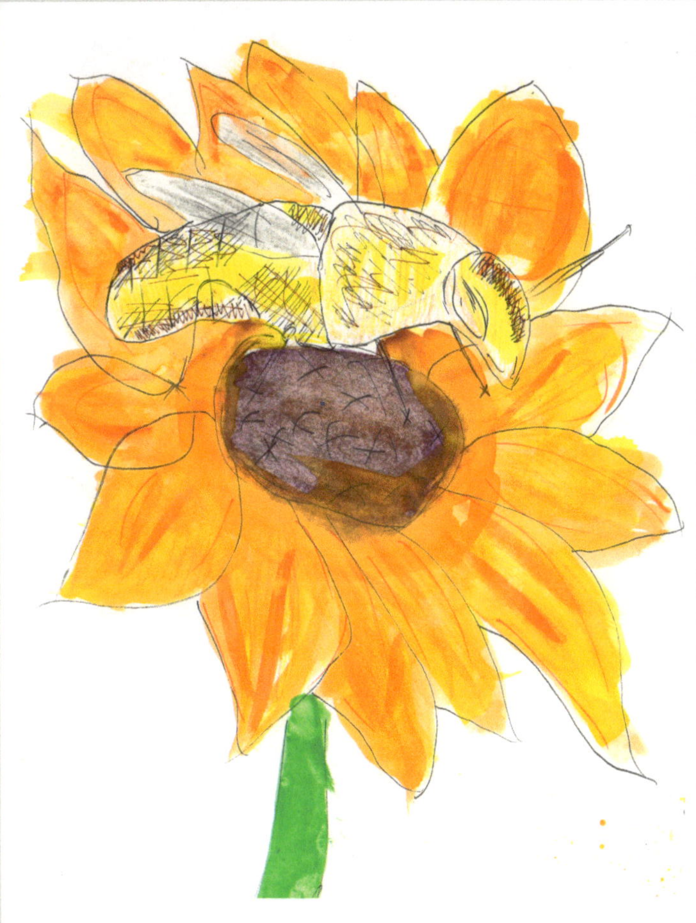
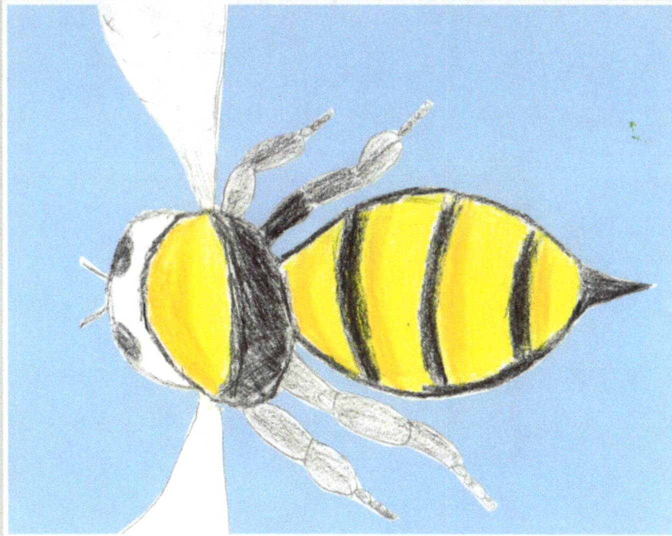
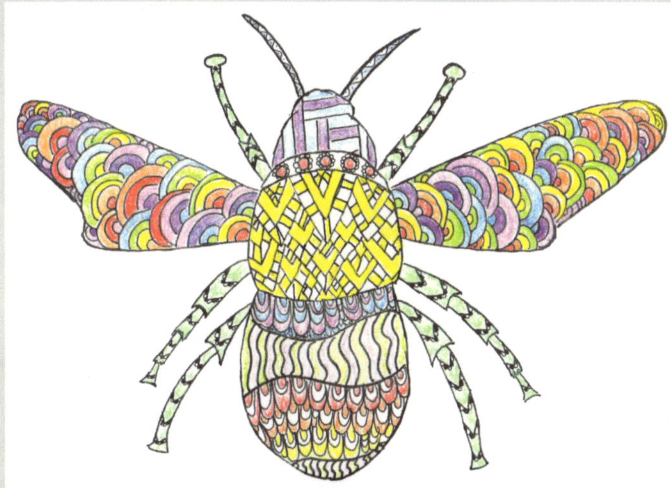
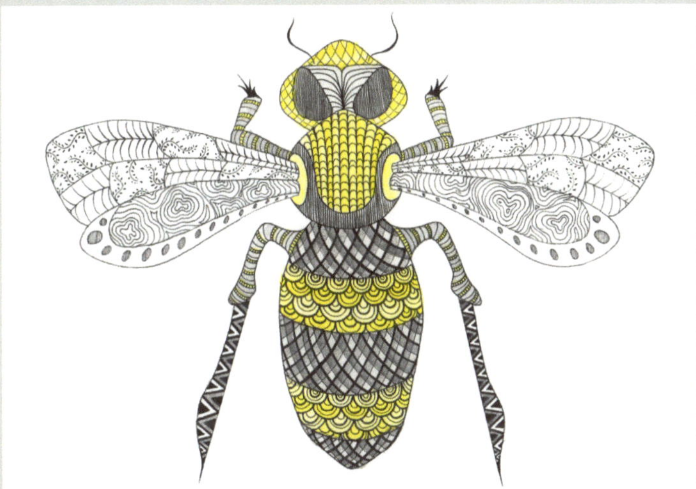
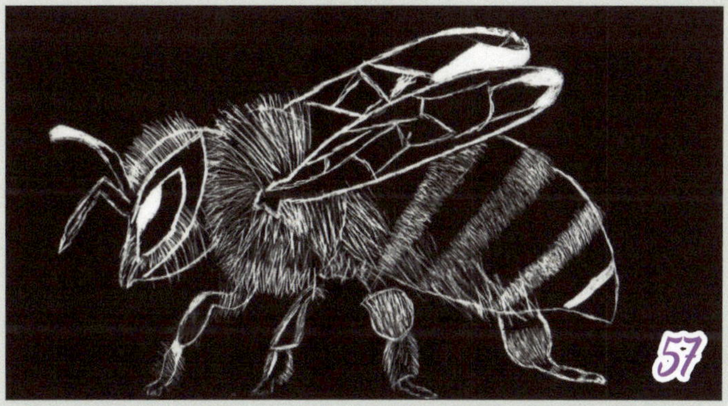

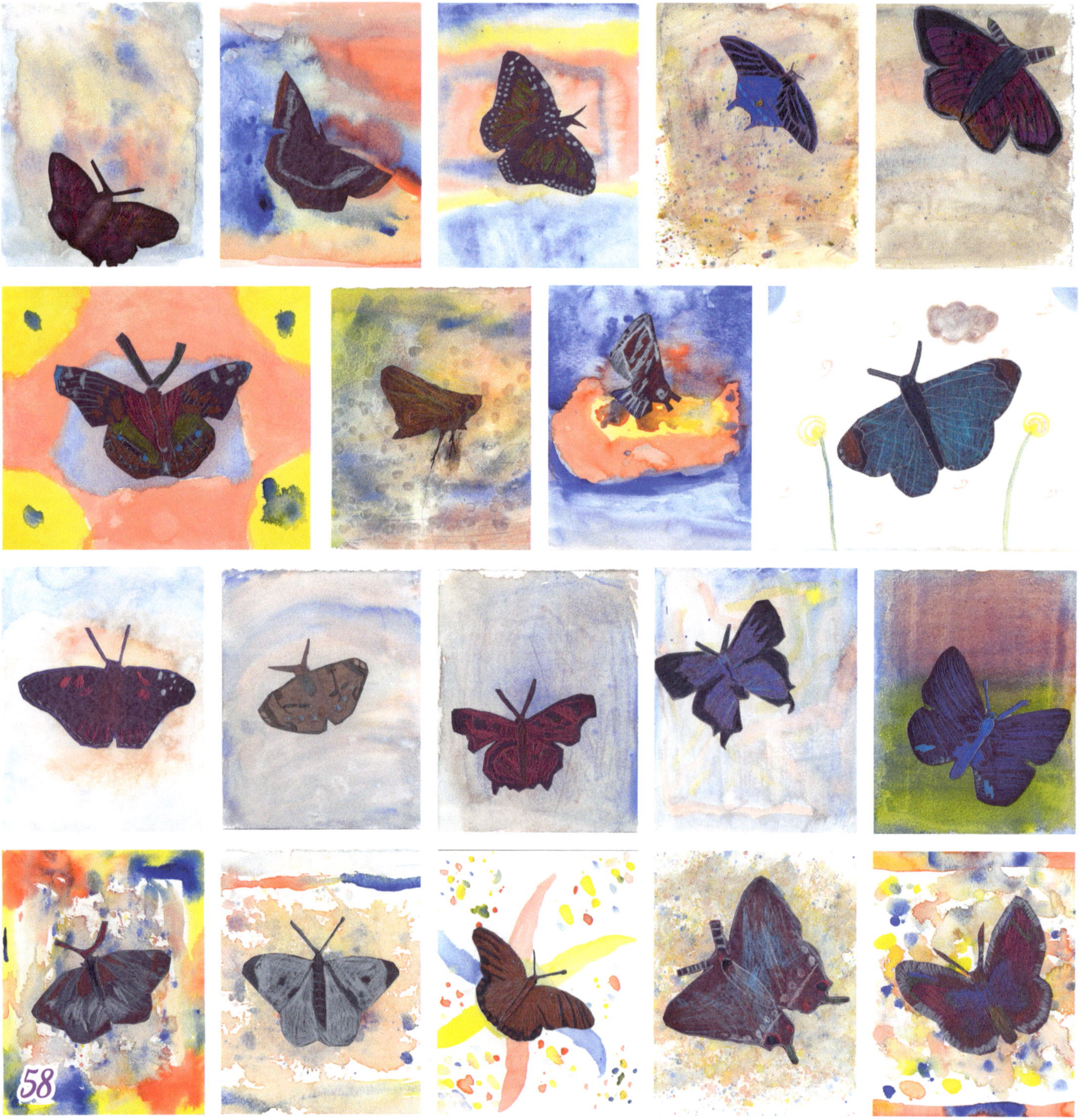

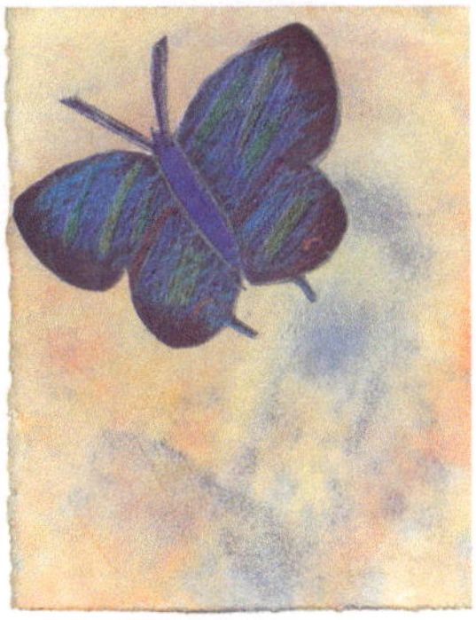
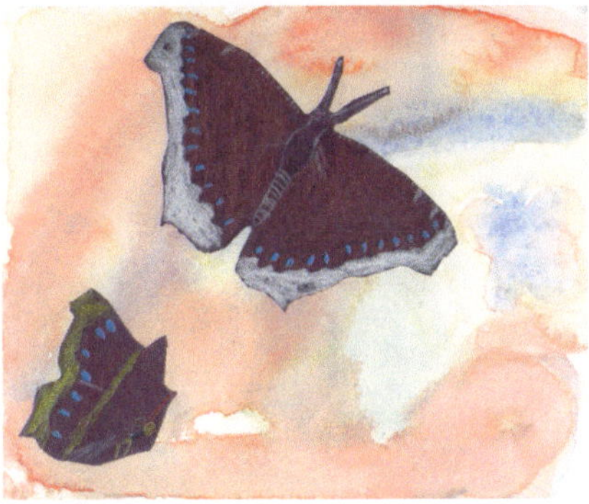
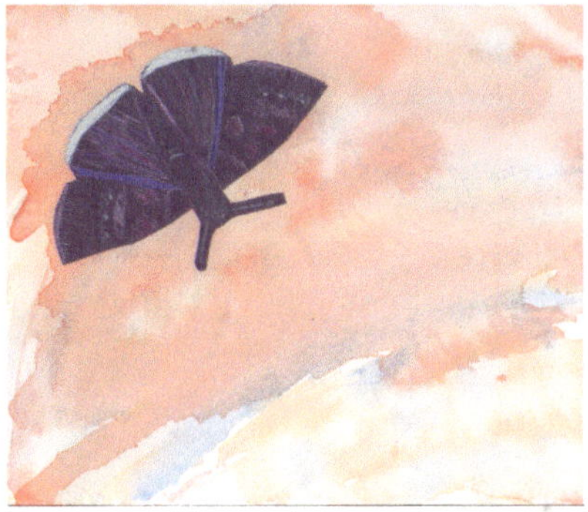
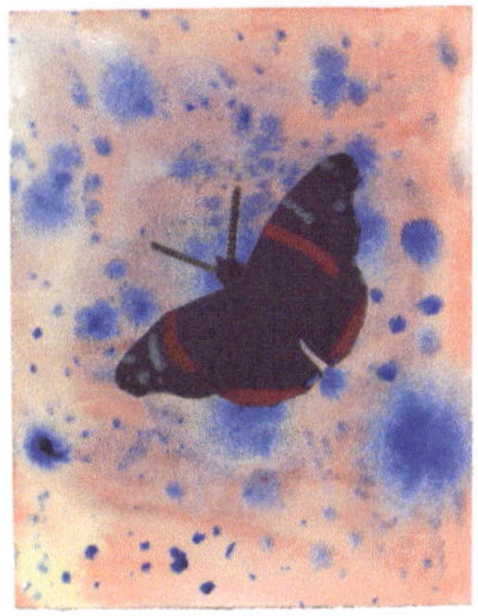
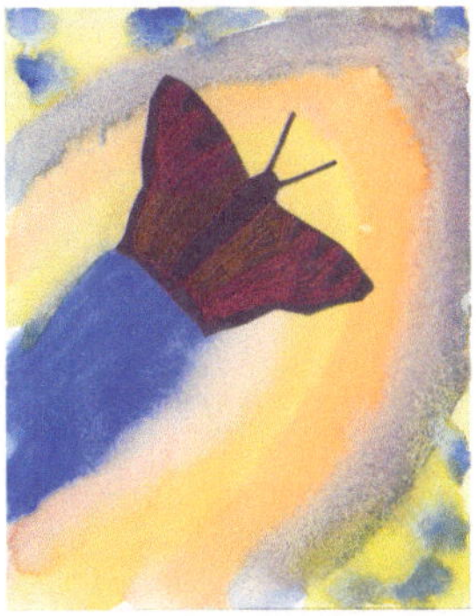
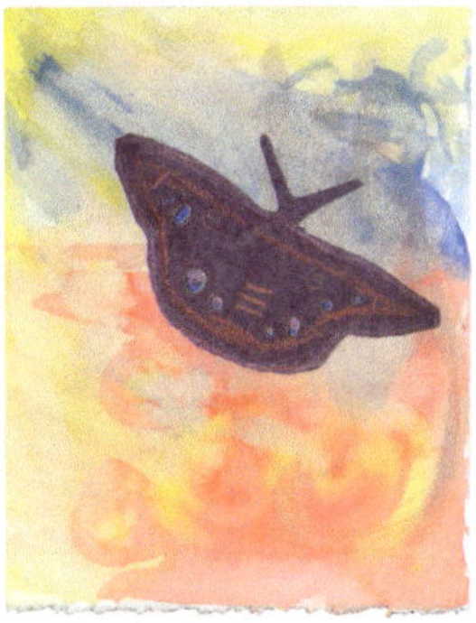
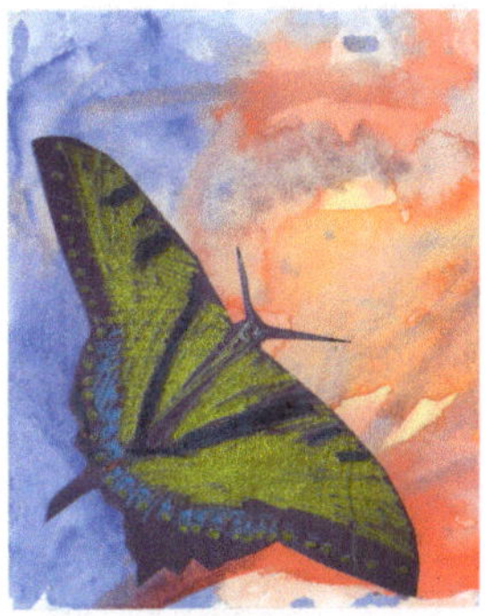
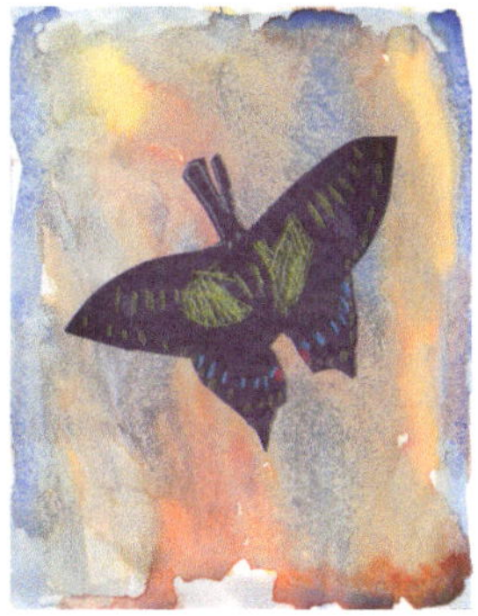
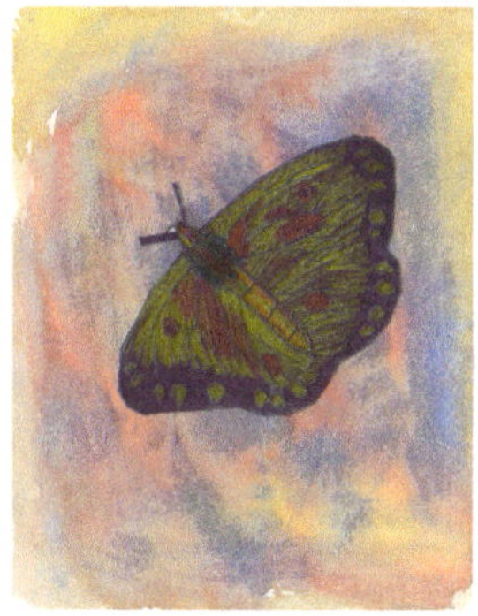
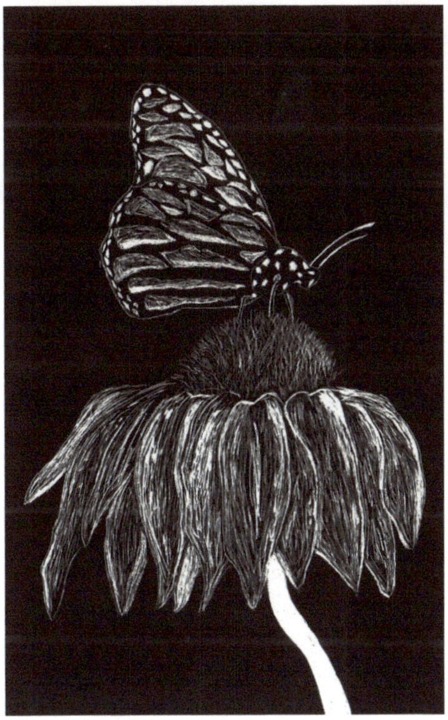
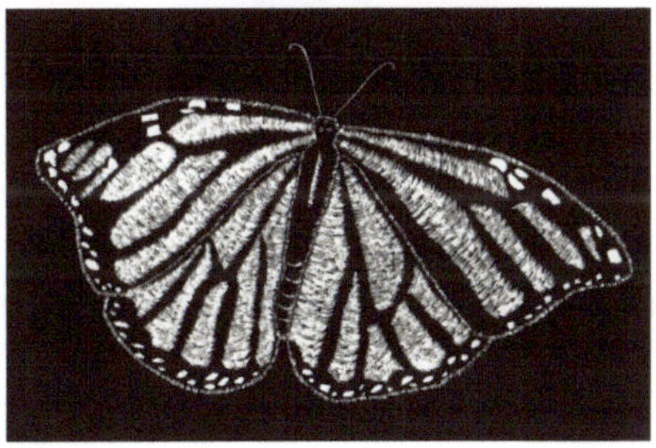
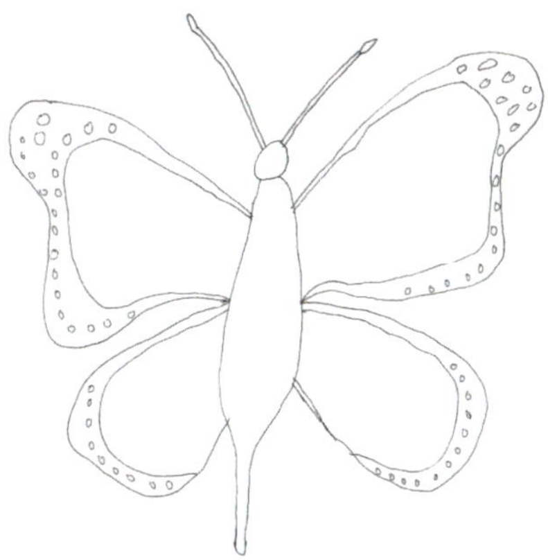
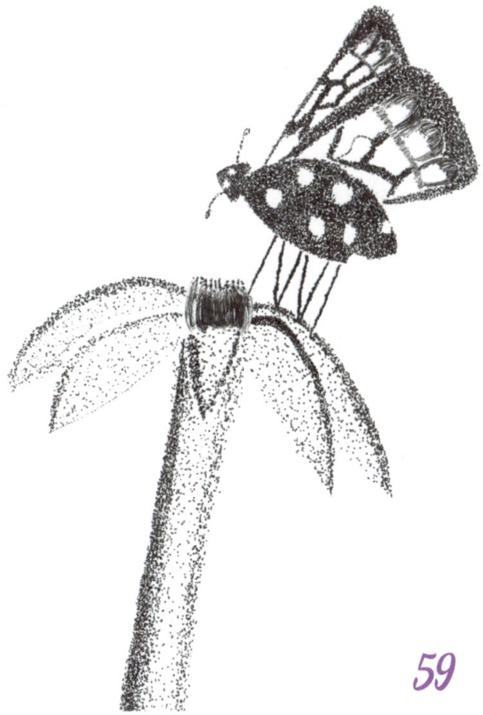

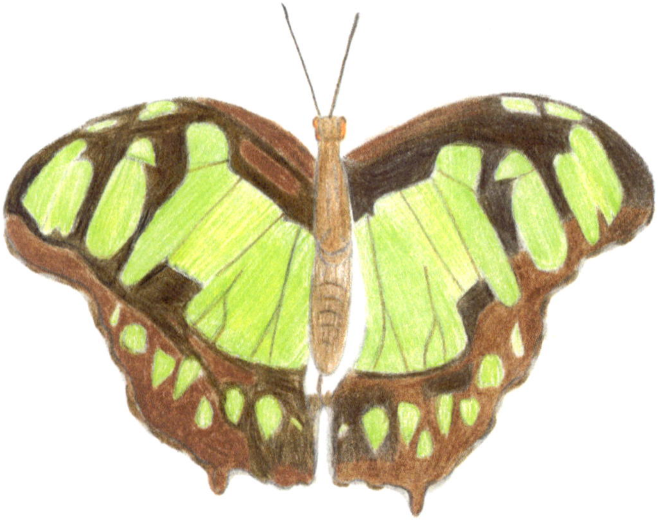

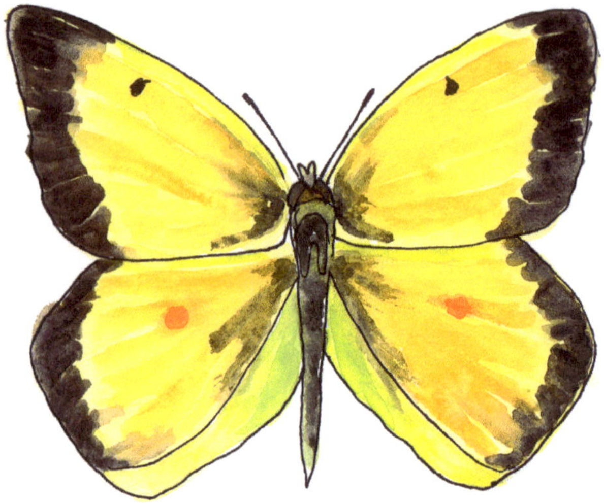

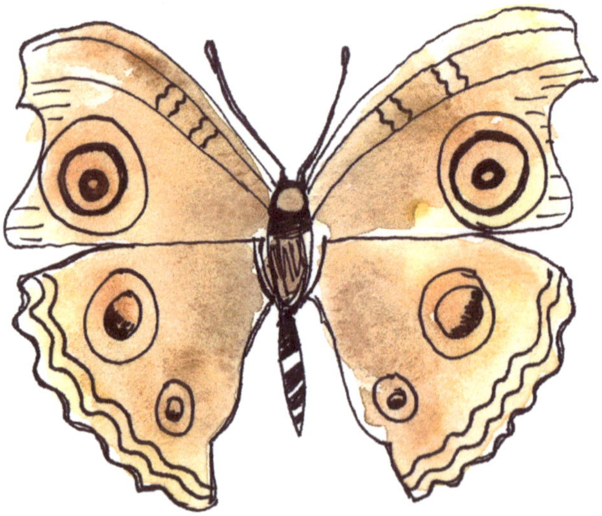

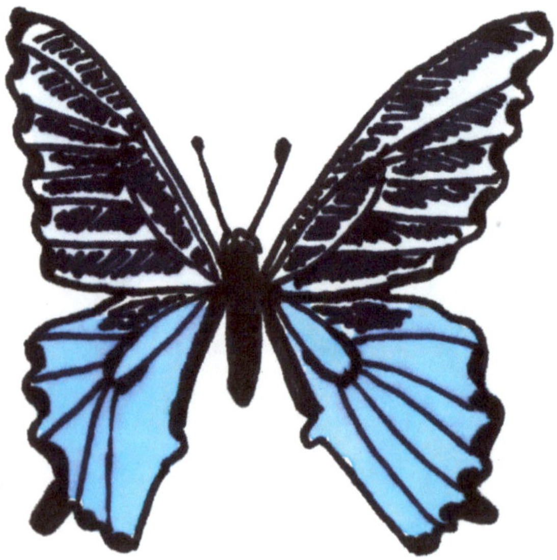

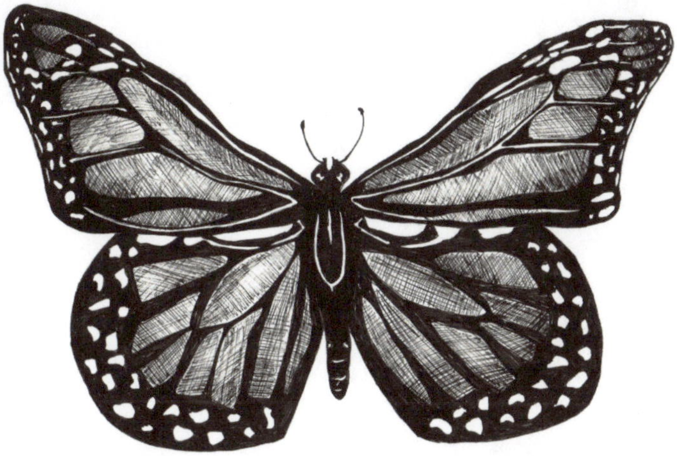

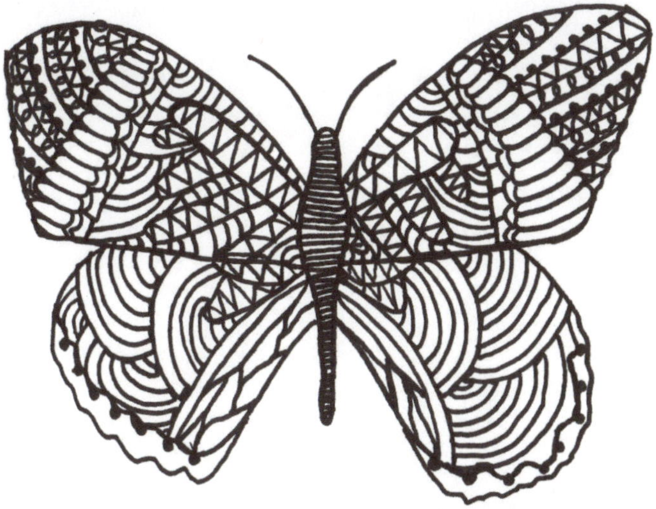

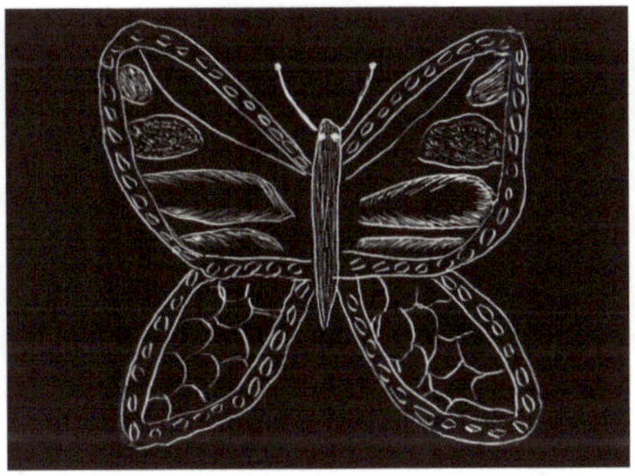

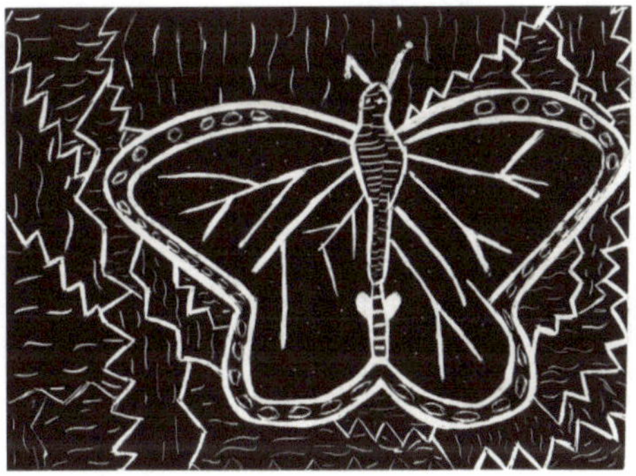

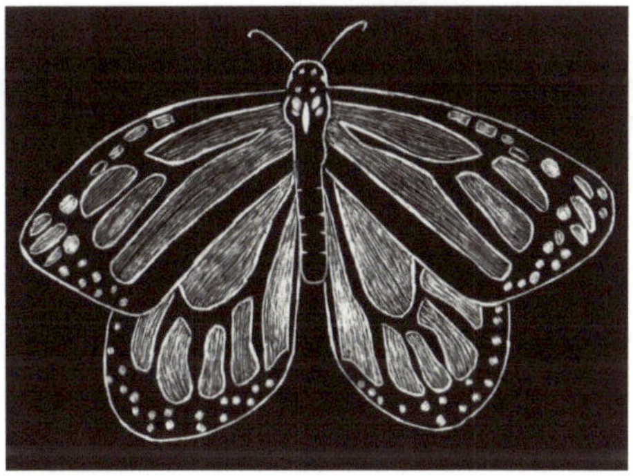

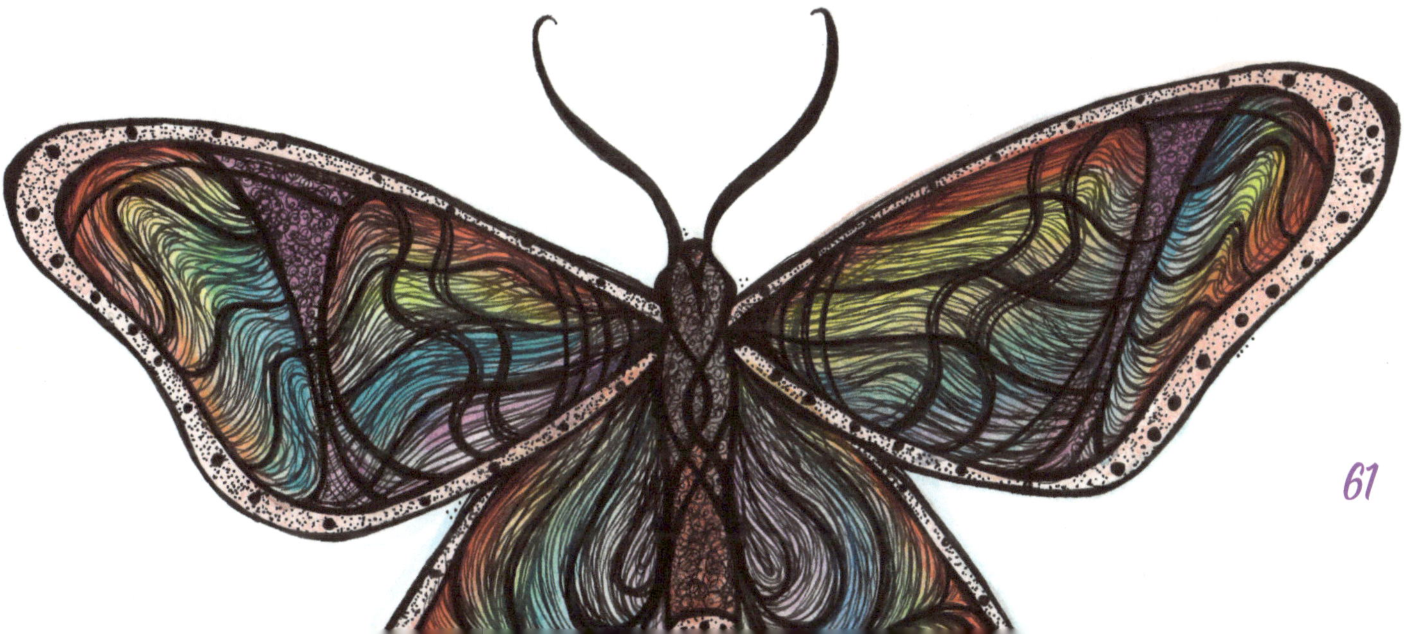

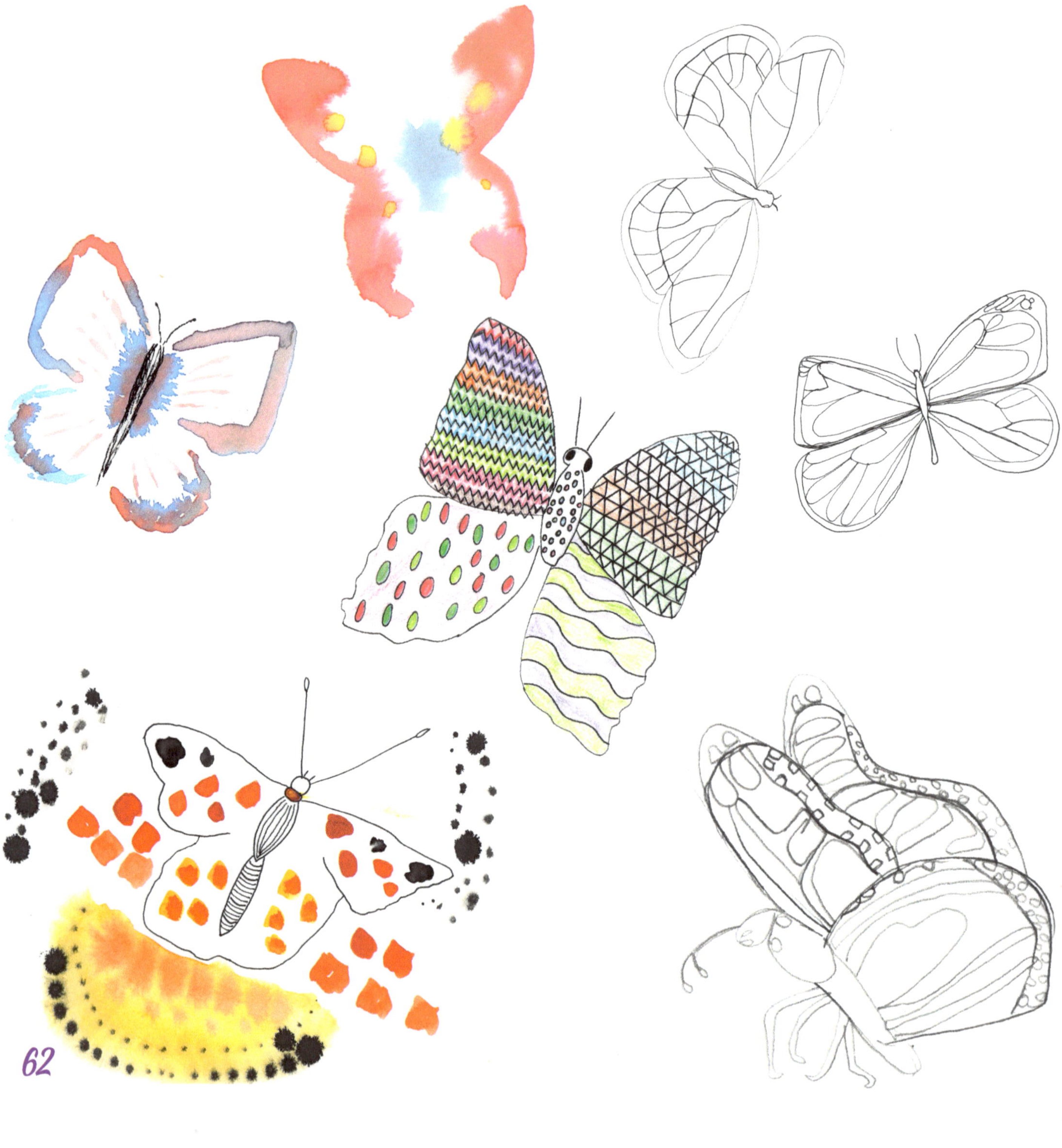

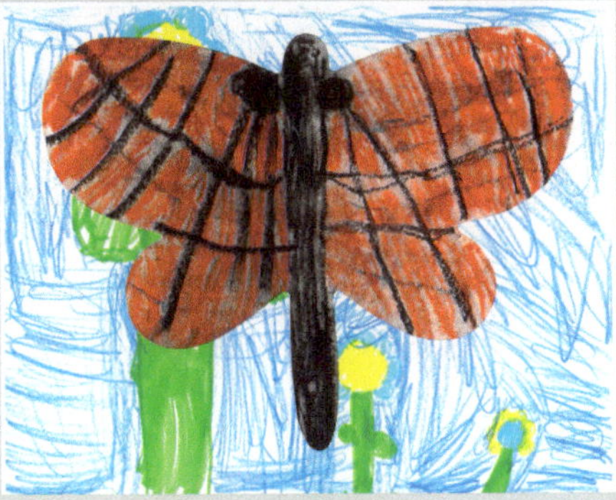
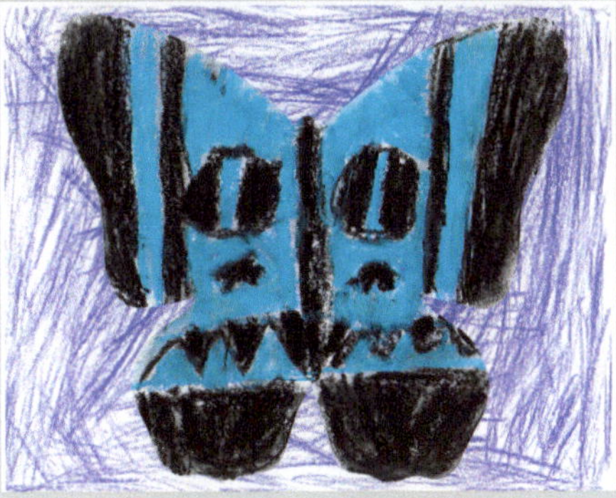
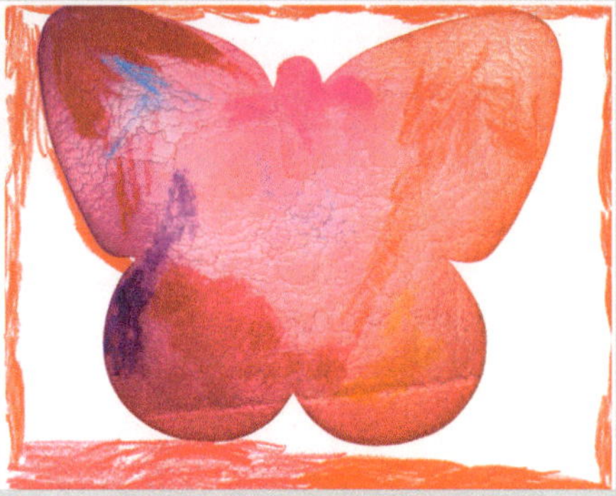
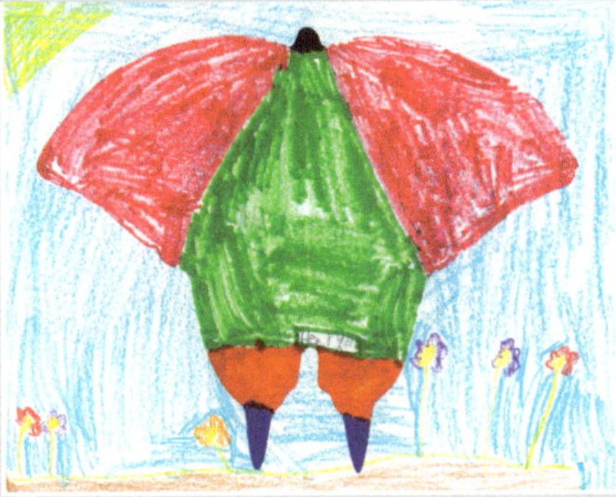
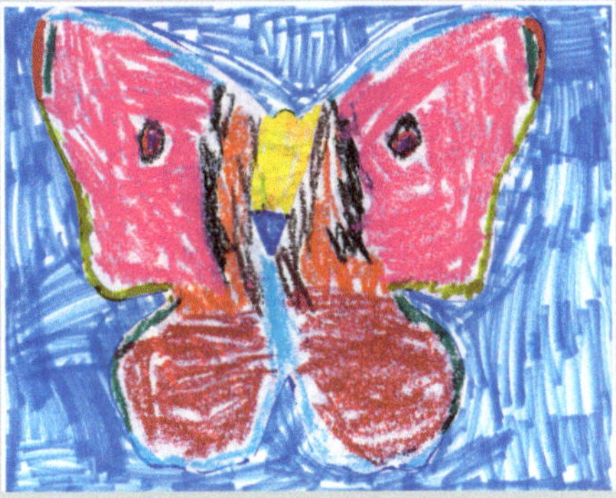
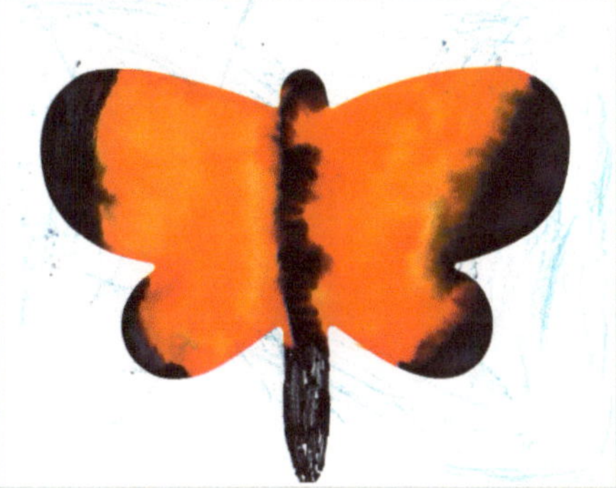
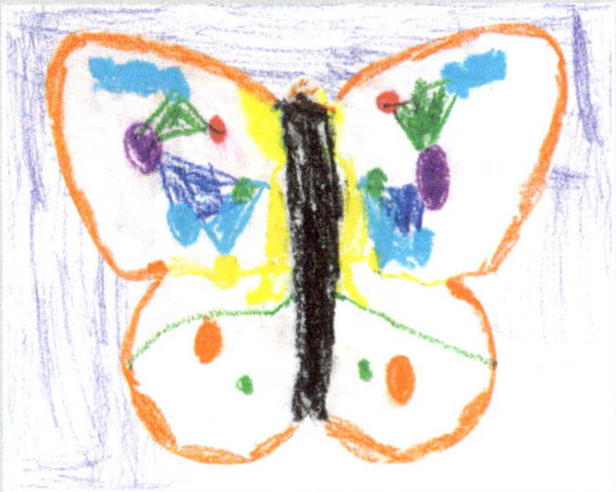
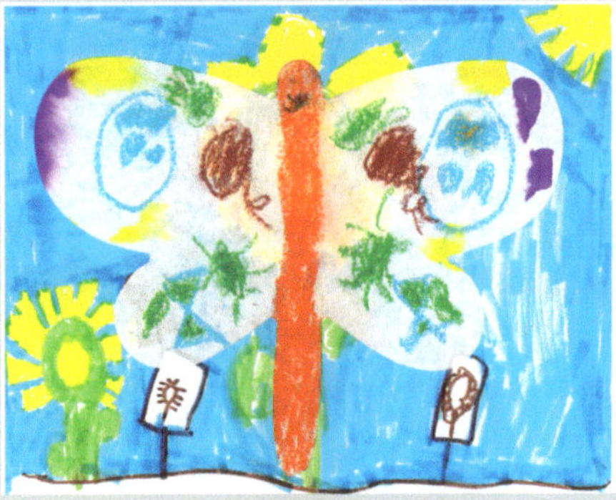
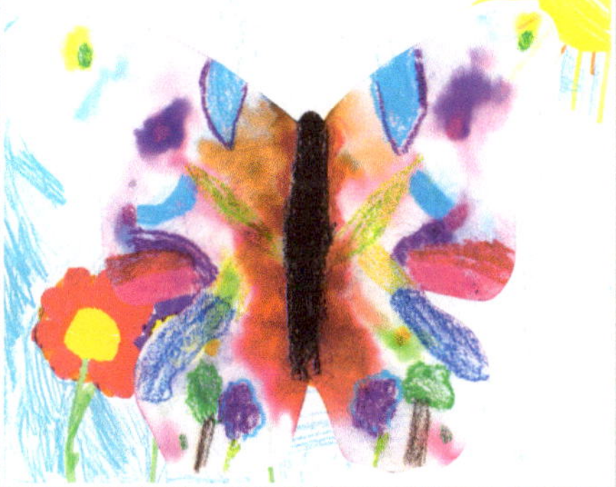
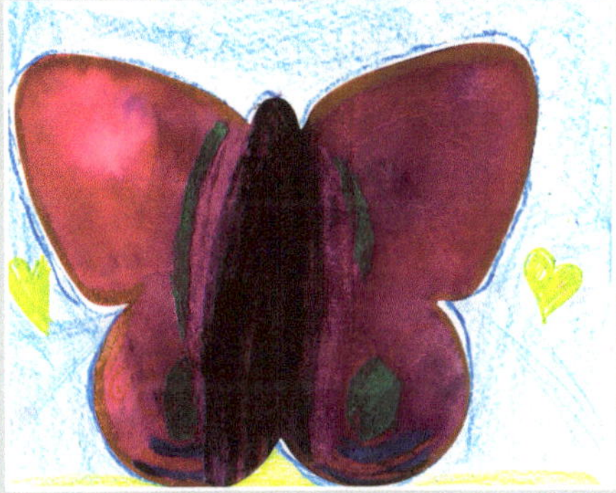
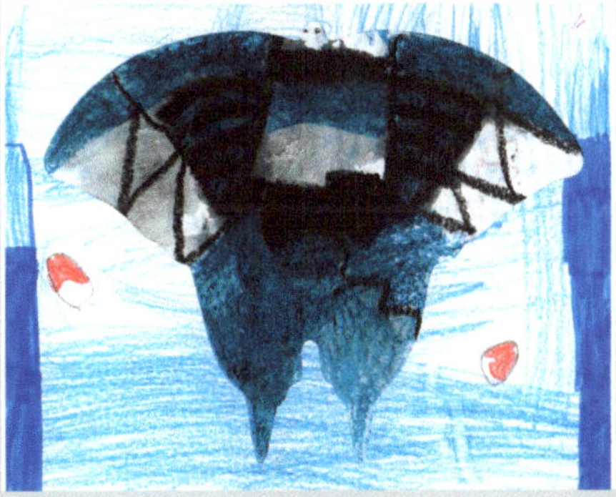
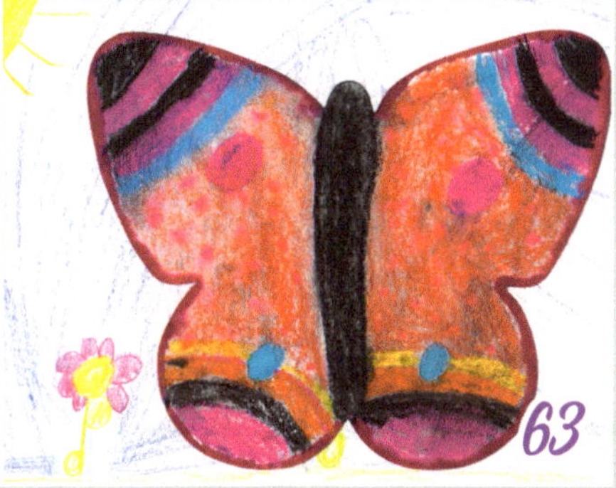

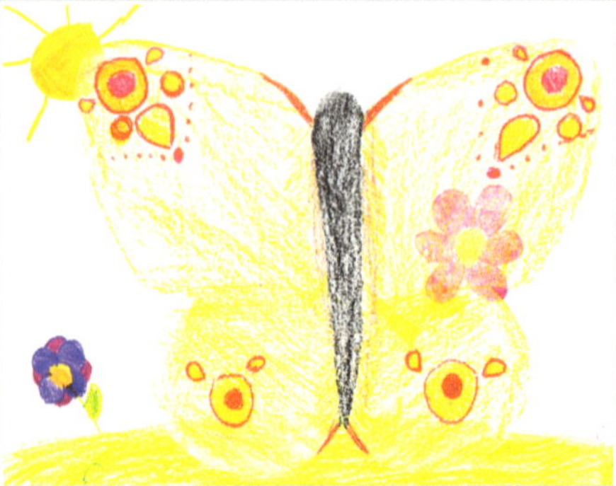
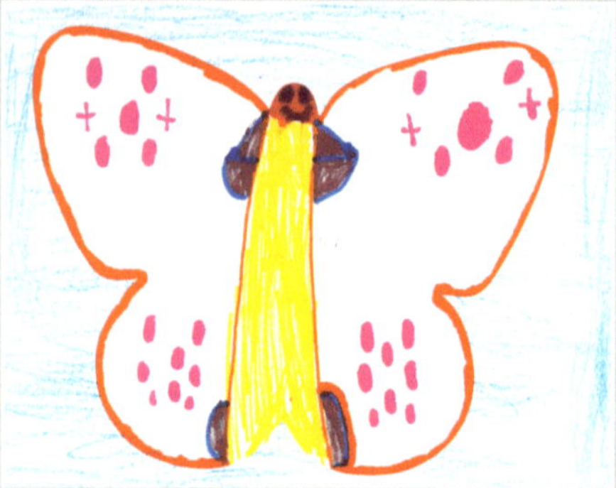
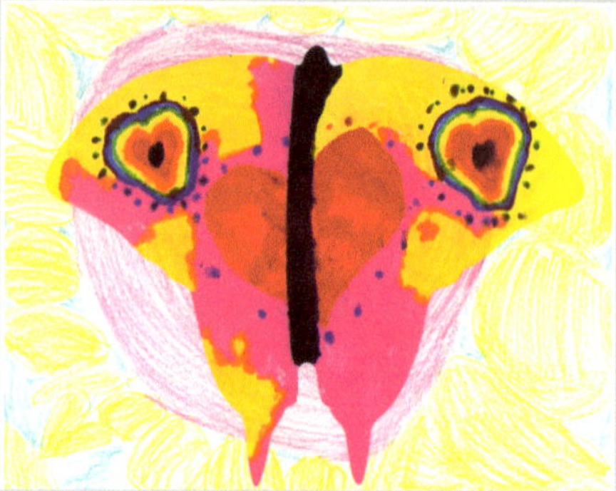
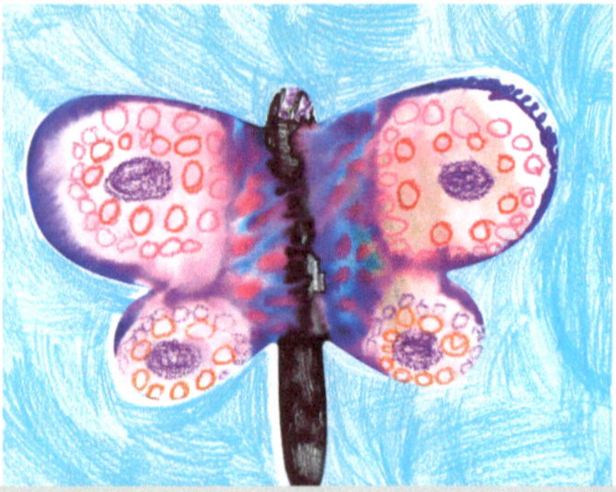
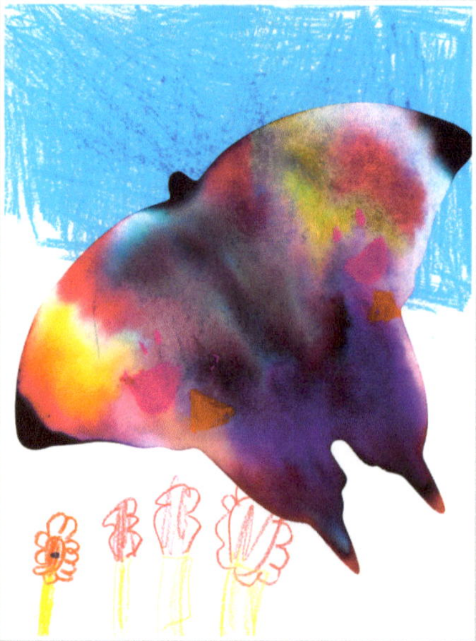
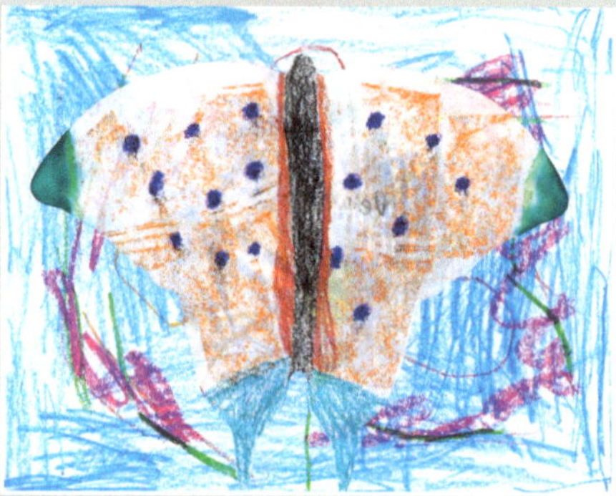
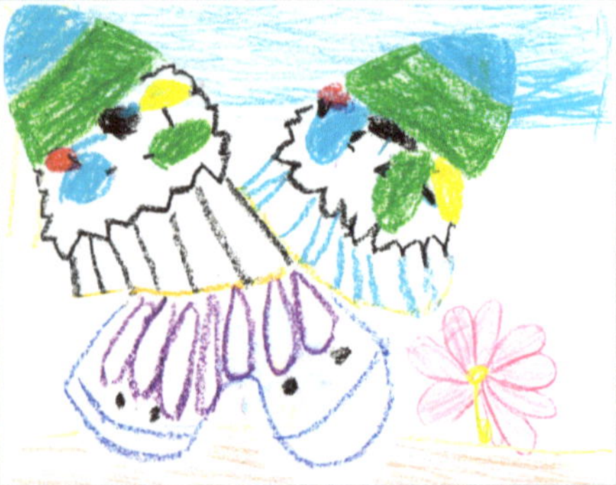
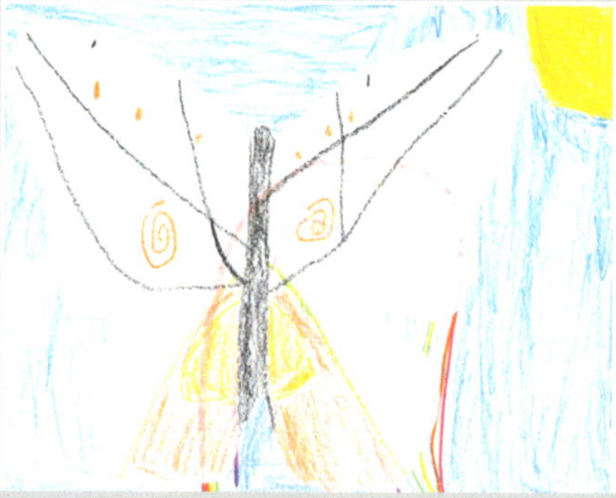
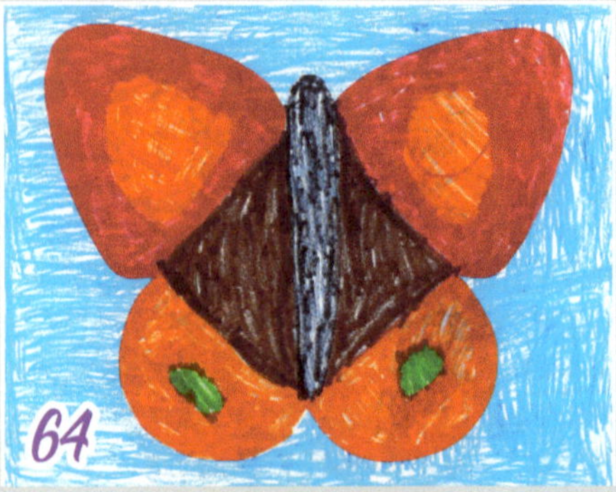
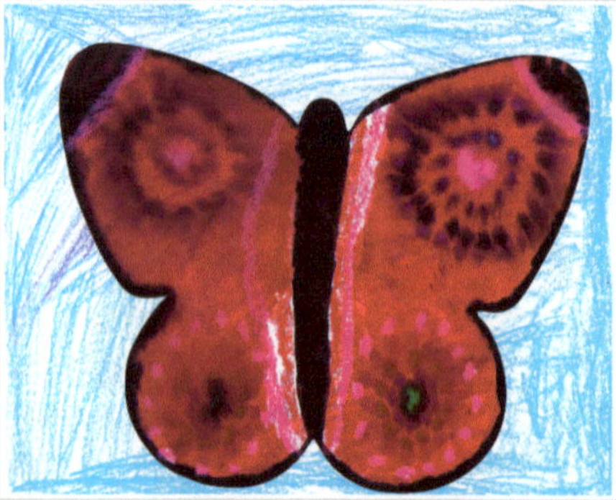
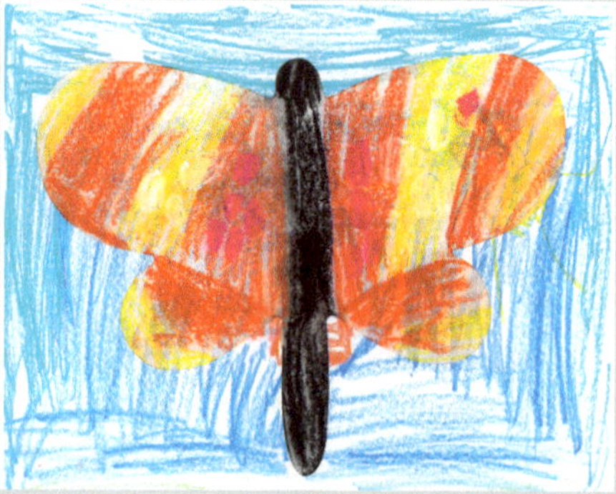

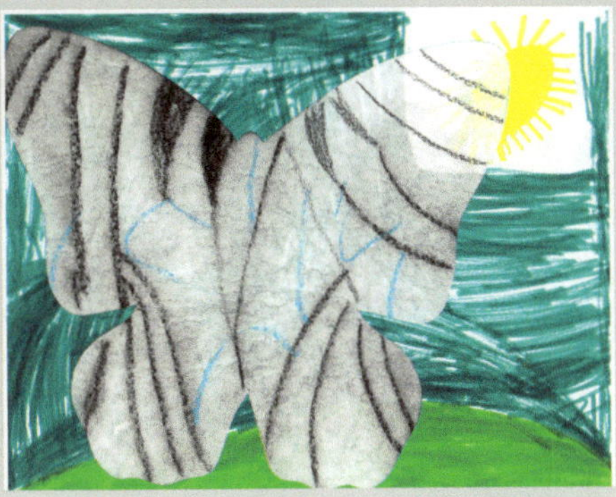
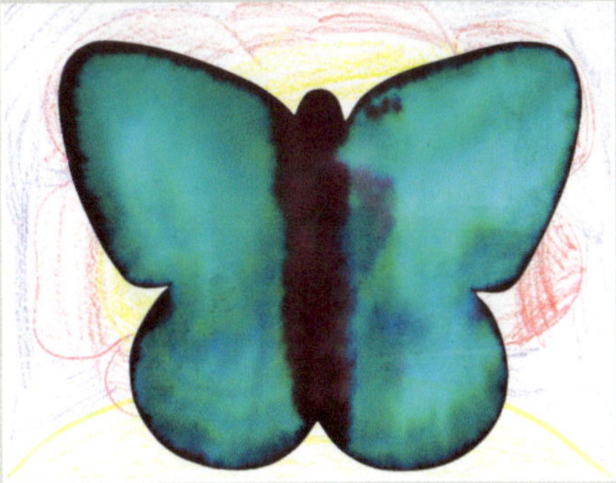
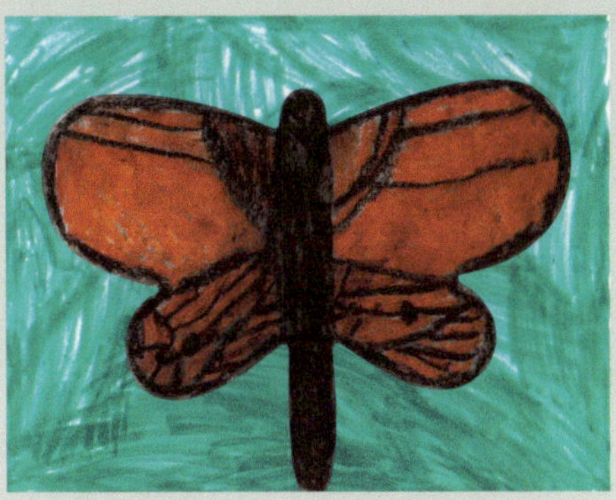
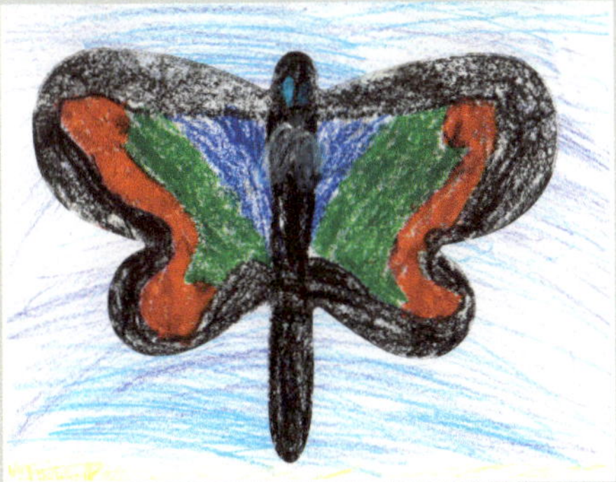
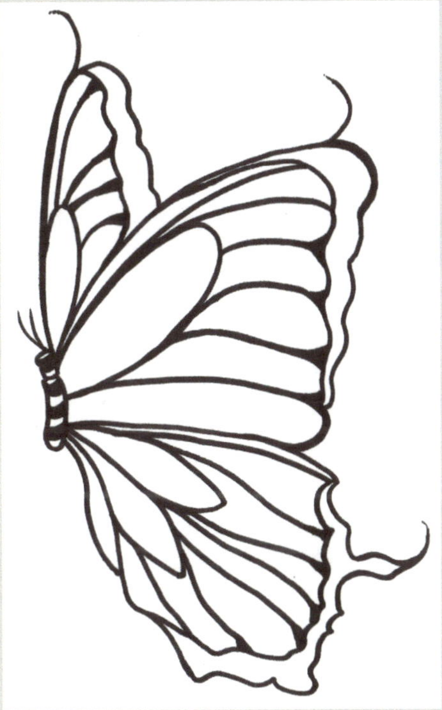
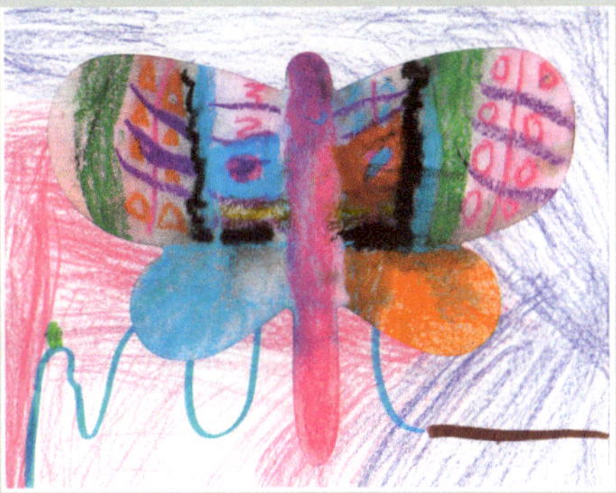
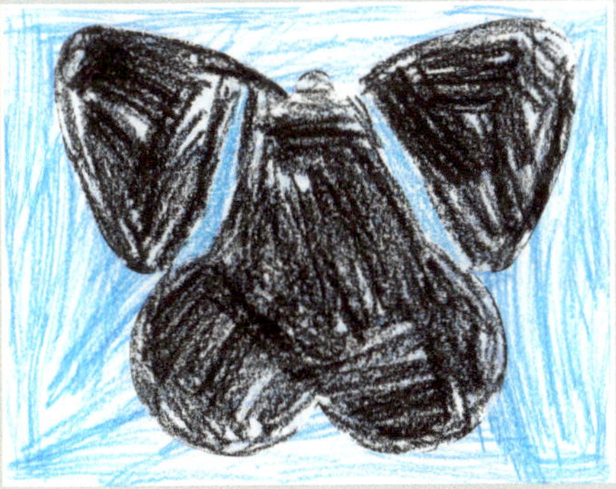
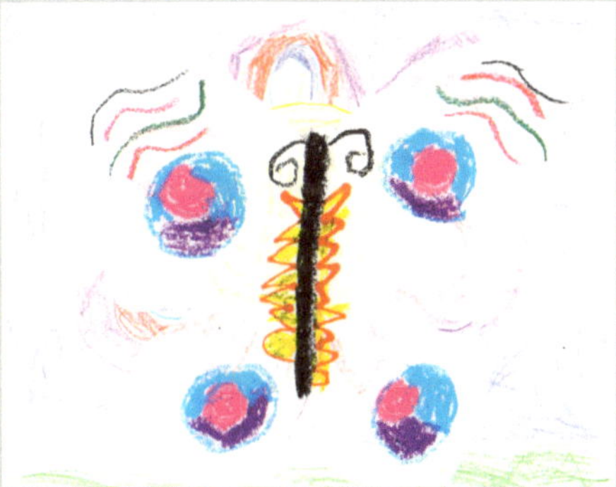
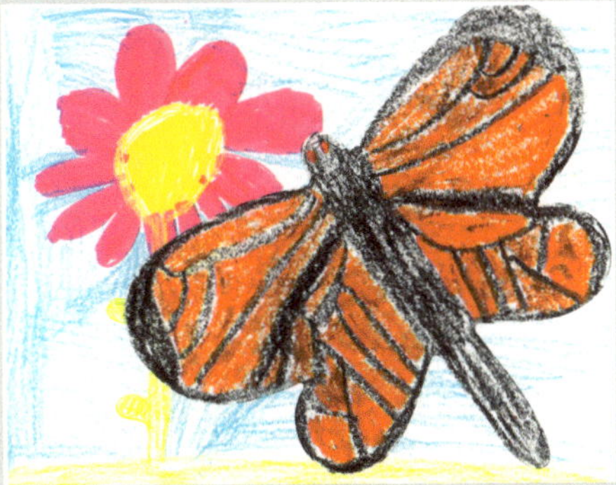
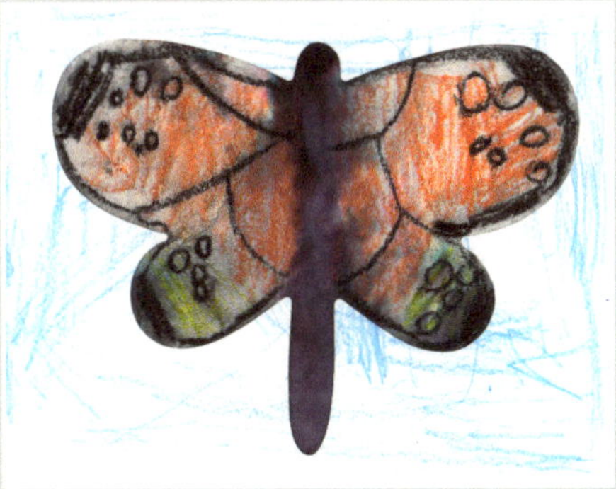

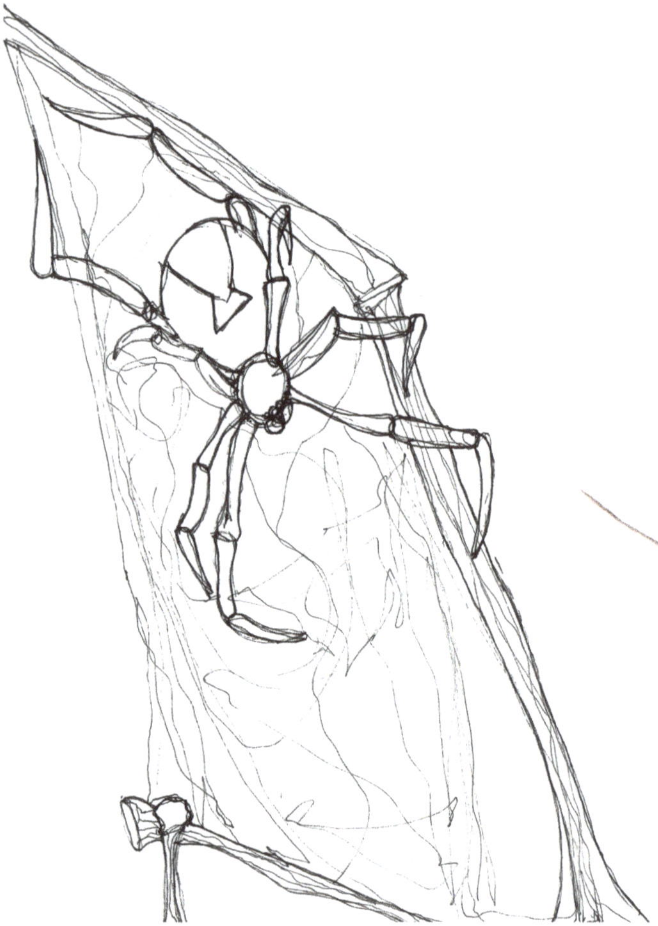

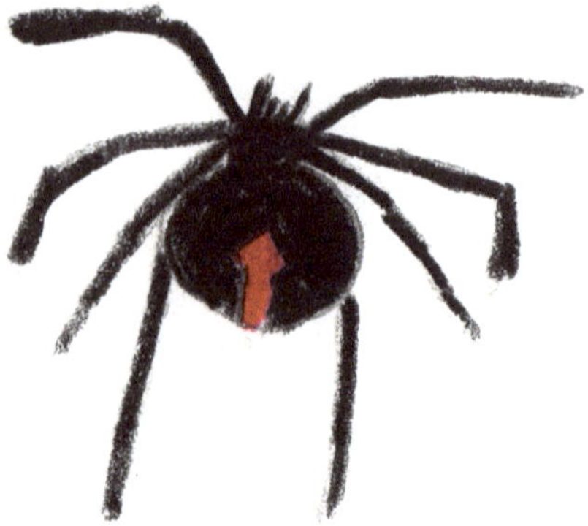

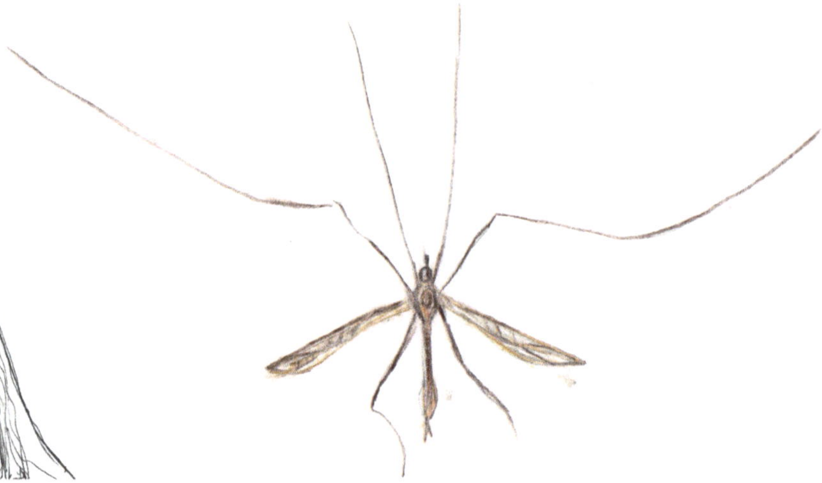

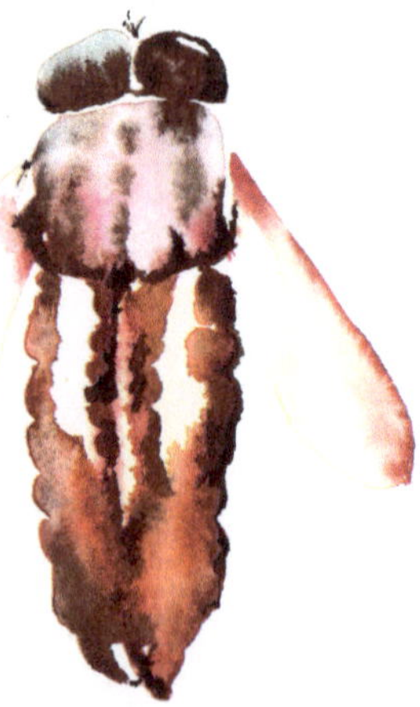

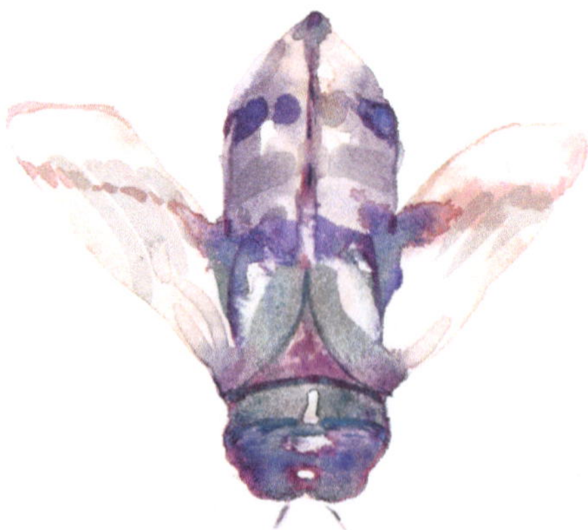

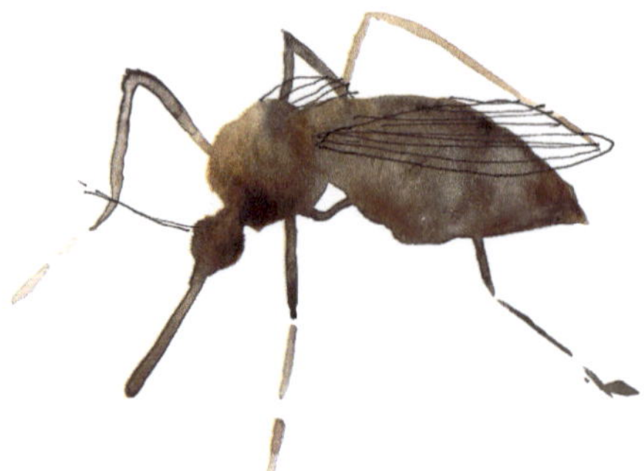

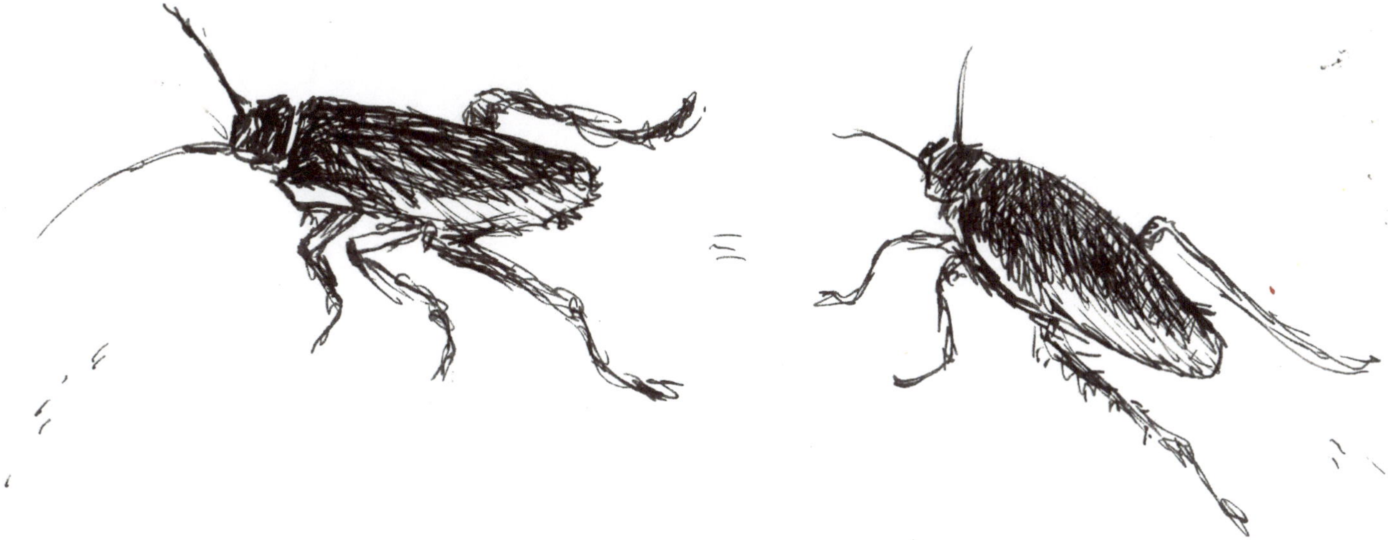

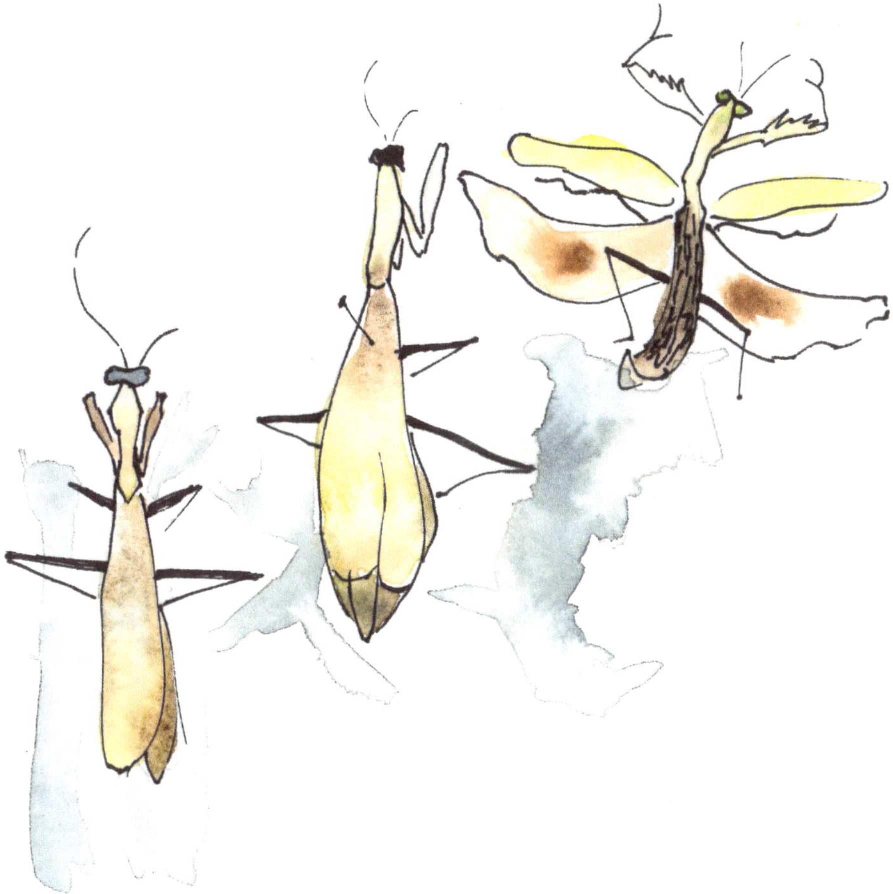

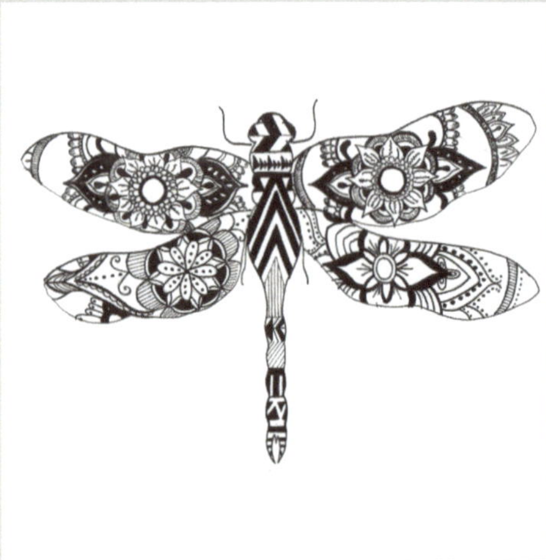
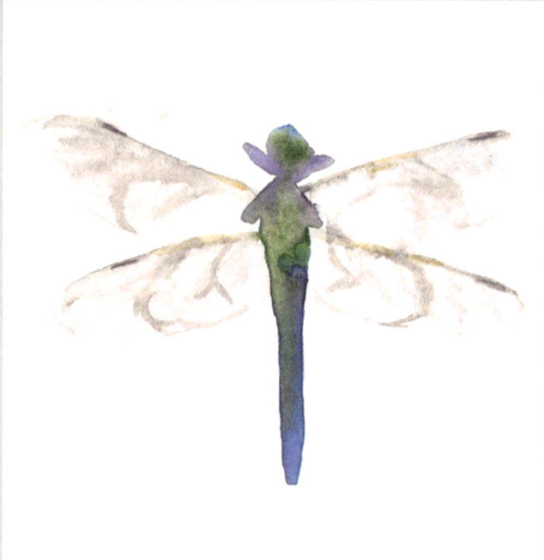
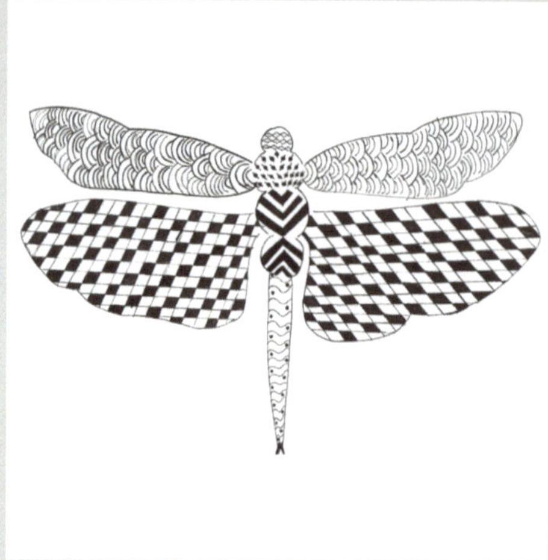
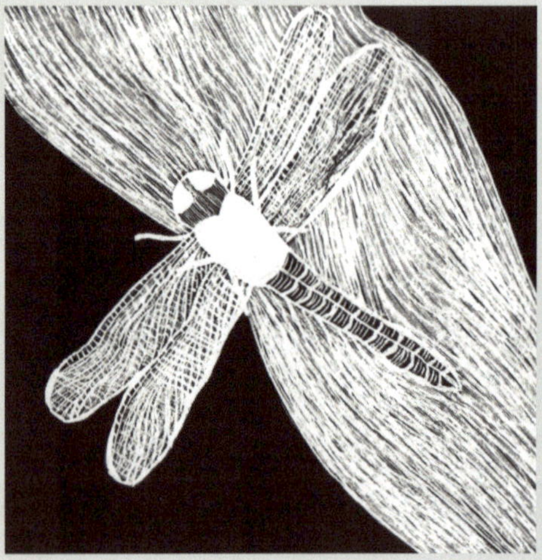
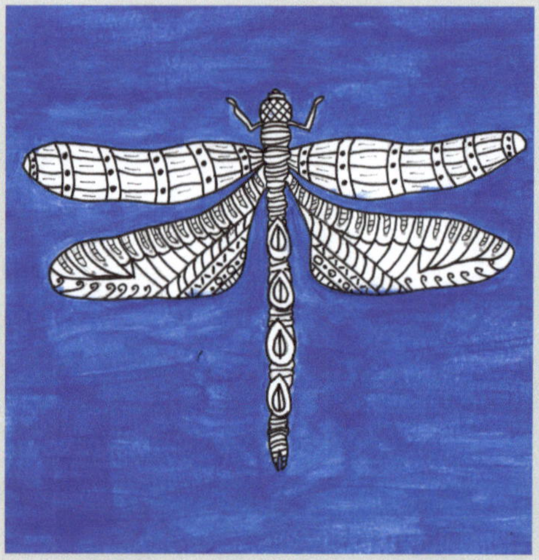
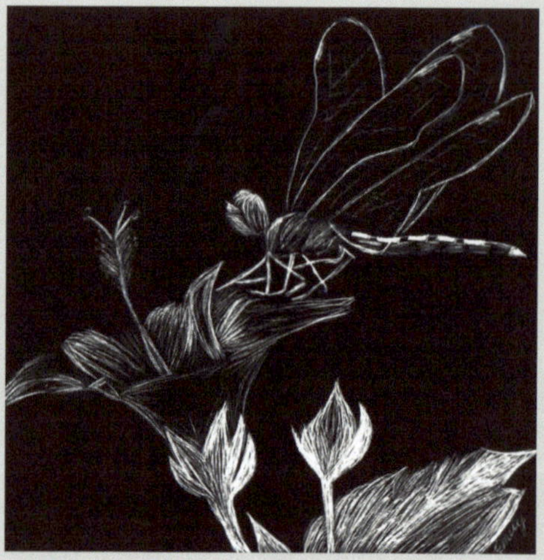
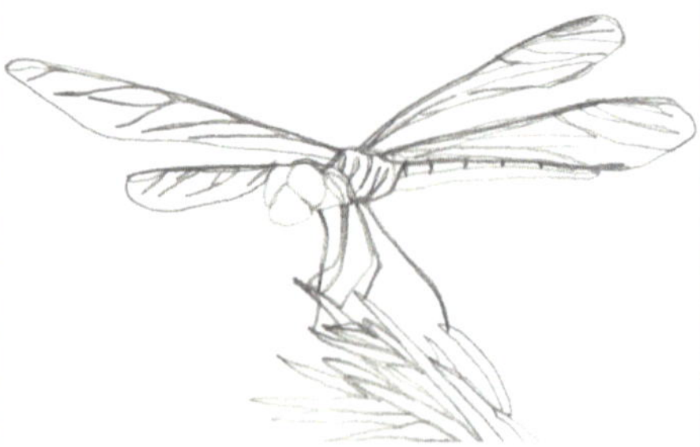
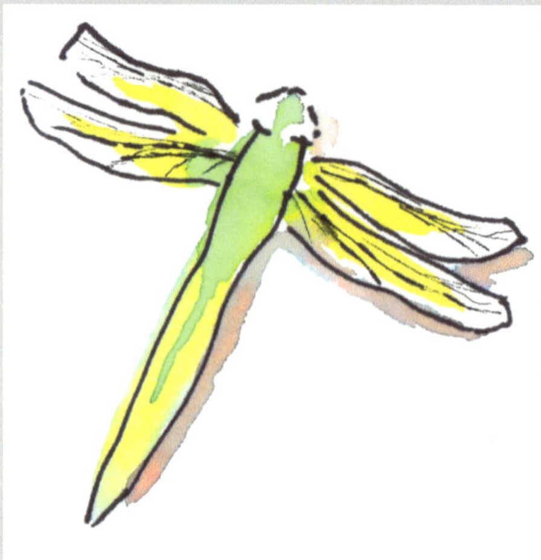
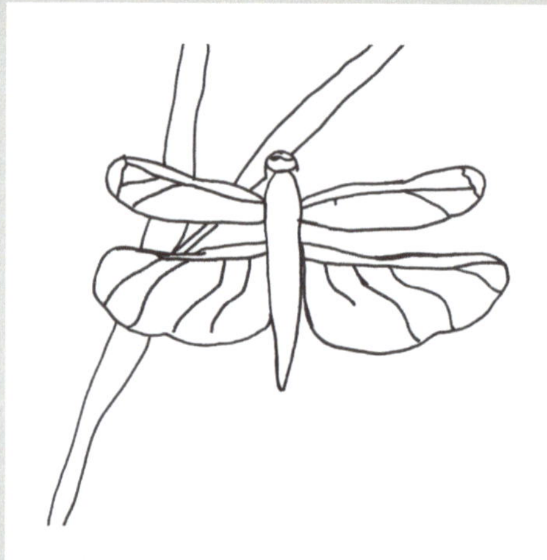

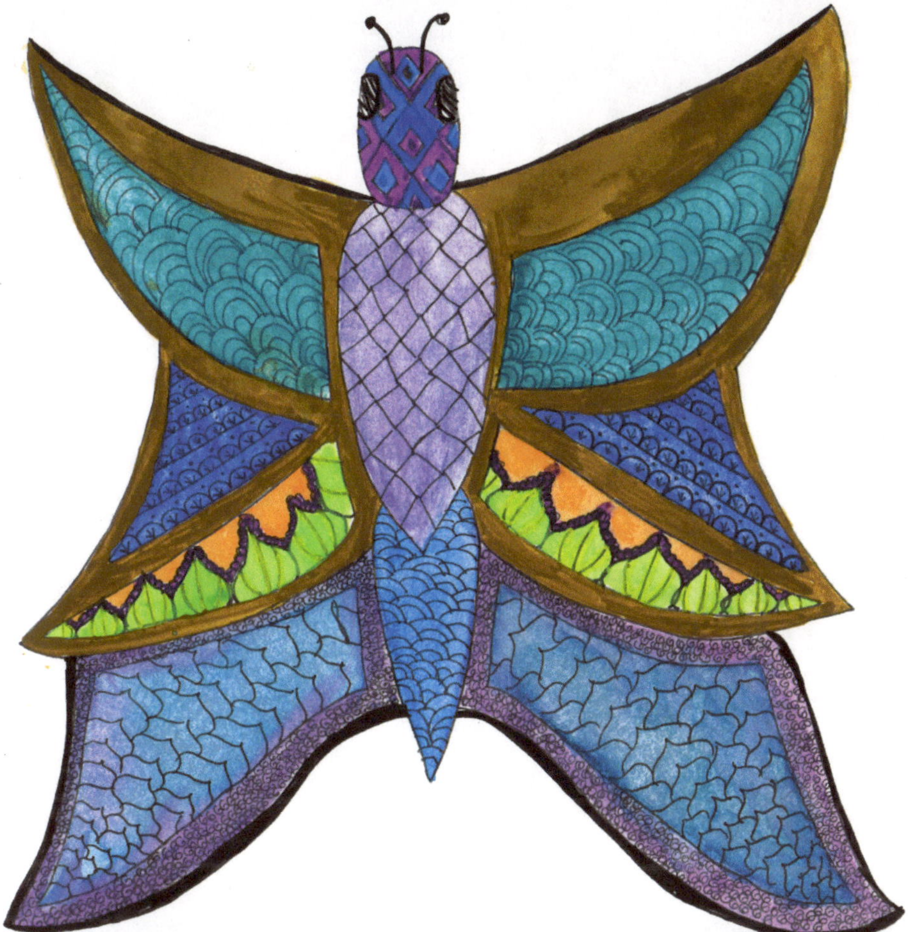

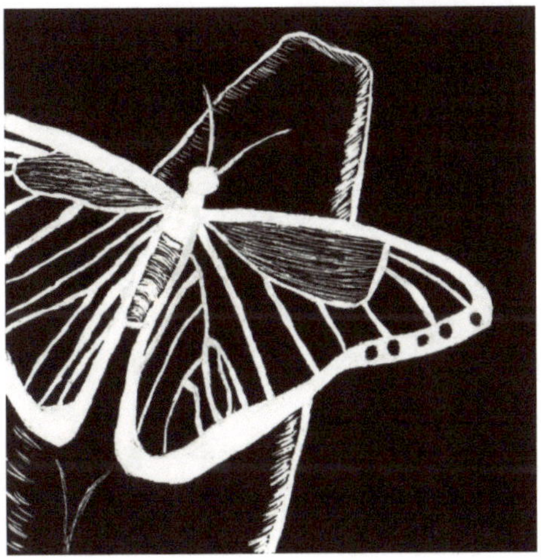

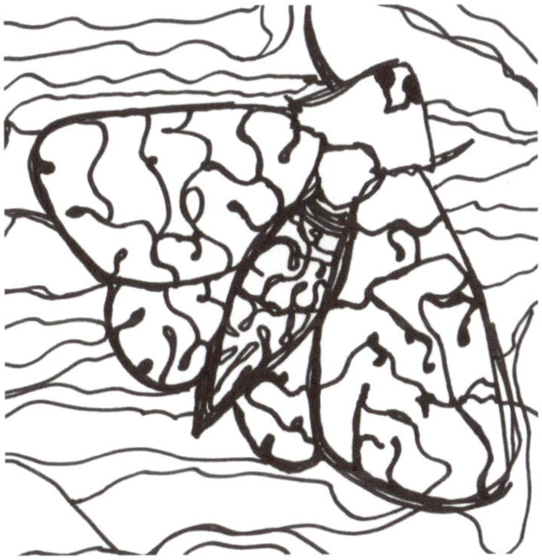

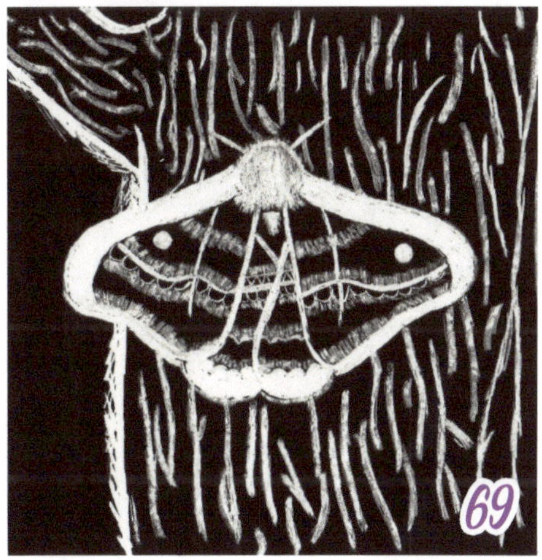

NTASY • BIRDS • BLACKBIRD • CANADA GEESE • CHICKEN • CR
OVE • DUCK • EAGLE • EGRET • FINCH • FLICKERS • HAWK • HER
UMMINGBIRD • MAGPIE • MOCKINGBIRD • OWL • SCRUB JAY • STAR
SWALLOW • TANAGER • TURKEY • TURKEY VULTURE • LANDSCA
LOUDS • DELTA BREEZE • FLATNESS • FOG • NORTH WIND • PU
REEK • SUNSET • PLANTS • ACORN • BERMUDA GRASS • ELDERBER
UCKEYE • BULBINE • CEANOTHUS • CENTURY PLANT • CHINESE PISTA
CHINESE TALLOW • COTTONWOOD • DAFFODIL • DATURA • DISCAR
HRISTMAS TREE • ELDERBERRY • FIG • FLOWERING QUINCE • FOX
RASS • GINKGO• HACKBERRY • HEUCHERA • IRIS • JASMINE • LEAF
LEMON • LUPINE • MILKWEED • MISTLETOE • MODESTO ASH • MUST
OLIVE TREE • ORNAMENTAL PEAR • PENSTEMON • PERSIMMON •
POMEGRANATE • POPPY • PURPLE NEEDLE GRASS • RED HOT POK
EDBUD • REDWOOD • ROSEMARY • SALVIA • SOURGRASS • SUNFLOW
YCAMORE • TOMATO • TOYON • TULE • VALLEY OAK • WALNUT • FISH
EPTILES • FROG • SALMON • TURTLE • MAMMALS • CAT • COW • COYO
OG • HORSE • OTTER • OPOSSUM • RABBIT • RACCOON • RAT • SQUIRR
NSECTS • BEE • BLACK WIDOW SPIDER • BUTTERFLY • COCKROACH • CR
Y • DEERFLY • DRAGONFLY • FLY • MOSQUITO • MOTH • PRAYING MA

FANTASY

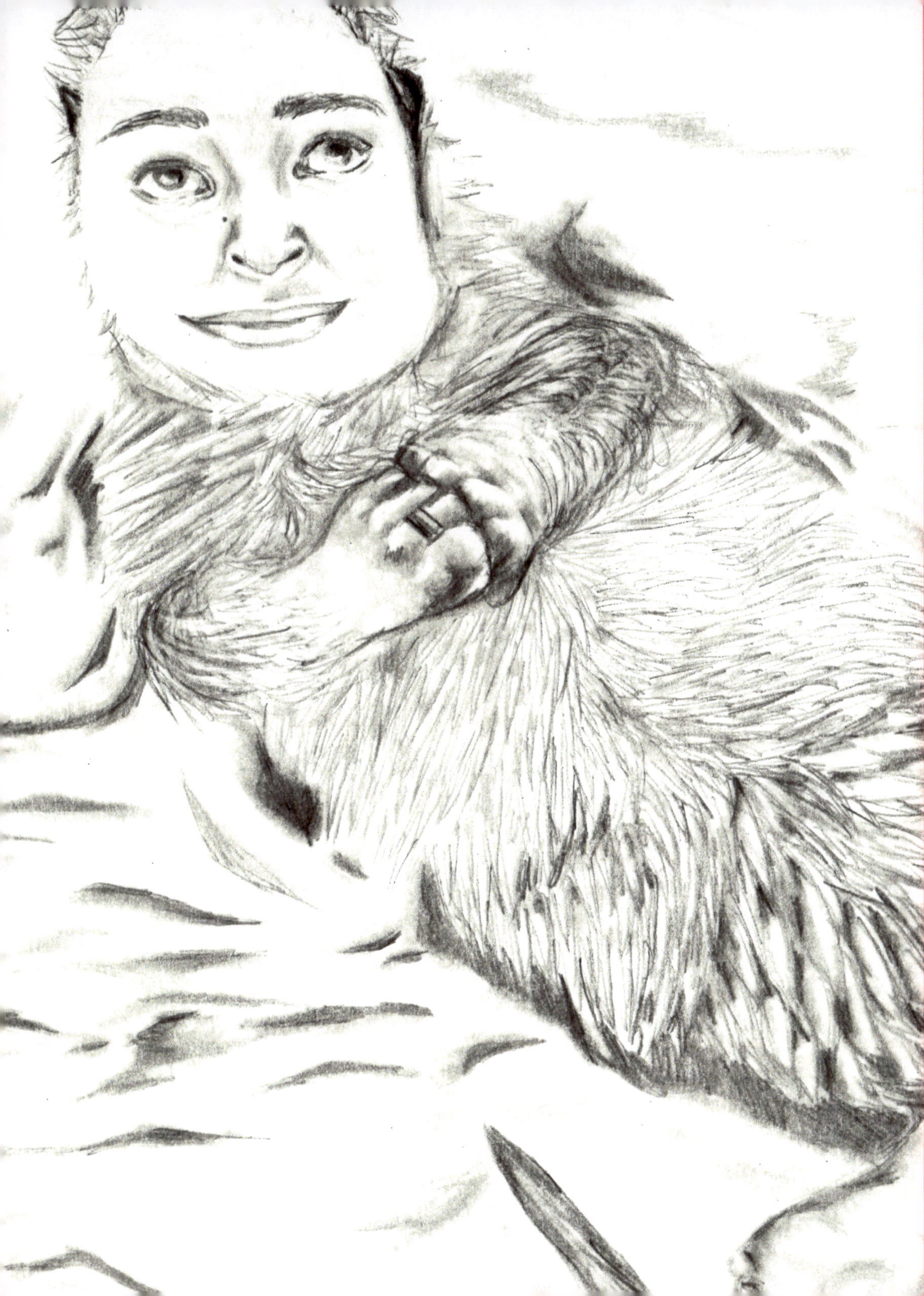

The following artists created the artwork in this section. The list is alphabetical by category and does not necessarily represent the order nor page number the artwork appears; please refer to the page numbers next to the category to locate where the art appears. Some teachers/schools did a class project (CP) for one category; all the artists involved in that project are listed in italics under the teacher and school name within the category. For other class projects the artist chose their category; their names appear under that category with a footnote indicating their teacher/school. If there is a number after an artist's name they provided more than one piece of art for that category.

HUMAN CHICKEN
(page 74)
Lee, Tammy - CP[1]

HUMAN COCKROACH
(page 73)
Lee, Tammy - CP[1]

HUMAN CROW
(page 75)
Clouser, Hannah - CP[1]

HUMAN DOG
(pages 76)
Gonzalez, Mayrin - CP[1]
Simpkins, Taylor - CP[1]

HUMAN GOOSE
(page 73)
Benush, Adithi - CP[1]

HUMAN HUMMINGBIRD
(page 75)
Garcia, Azalea - CP[1]
Mellberg, Grace - CP[1]

HUMAN IRIS
(page 77)
Murphy, Jenna - CP[1]

HUMAN OWL
(page 74)
Remme, Annika - CP[1]

HUMAN RED-WINGED BLACKBIRD
(page 75)
Carroll, Sarah, - CP[1]

HUMAN RIVER OTTER
(page 71)
Melavic, Megan - CP[1]

HUMAN SQUIRREL
(page 76)
Lake, Bella - CP[1]

HUMANOID OTTER
(page 111)
Seifert, Megan - CP[1]

[1]Student's Choice Class Project, Teacher: Melanie Bown, St. Francis High School • [2]Student's Choice Class Project, Teacher: Carrie Ceniseroz, St. James School

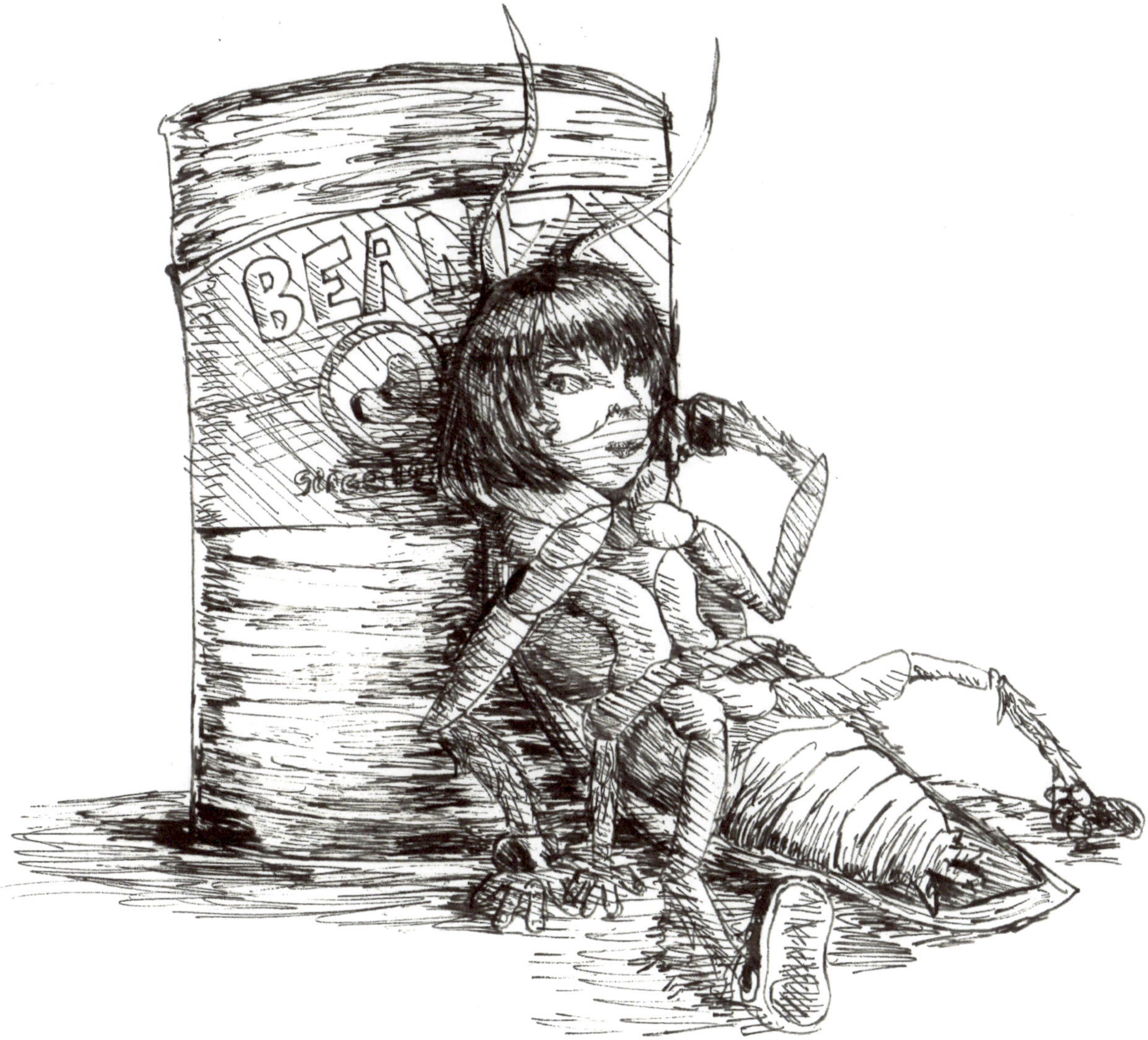

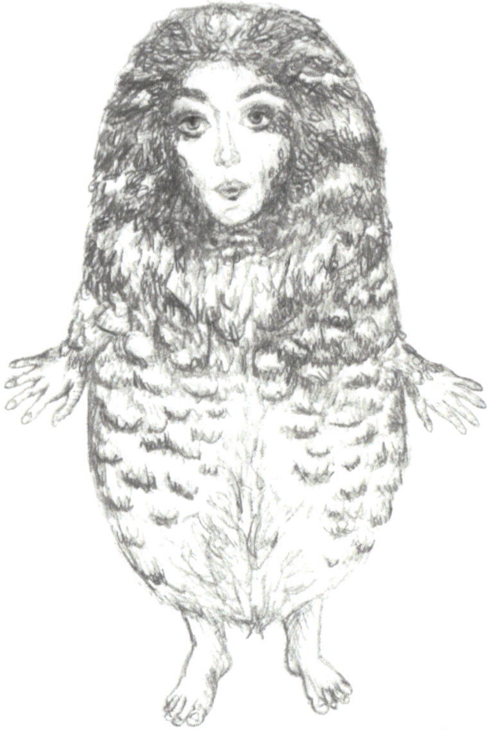

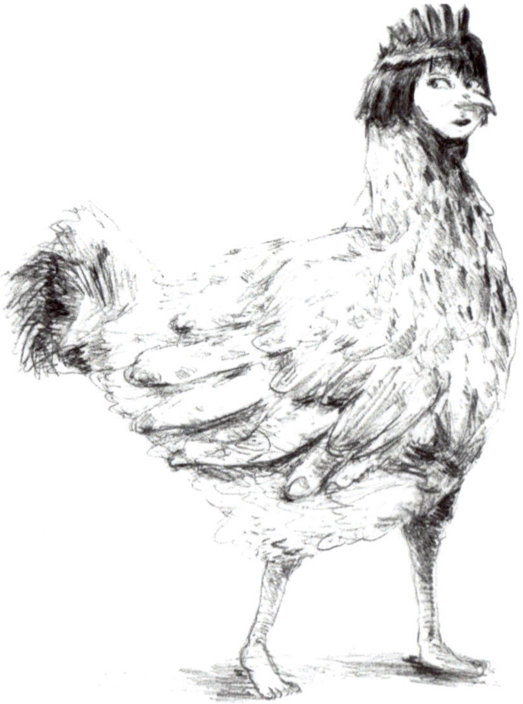

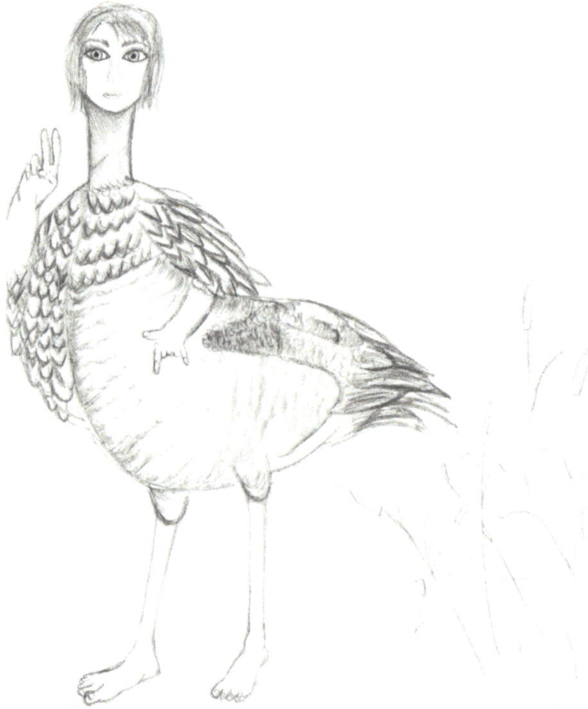

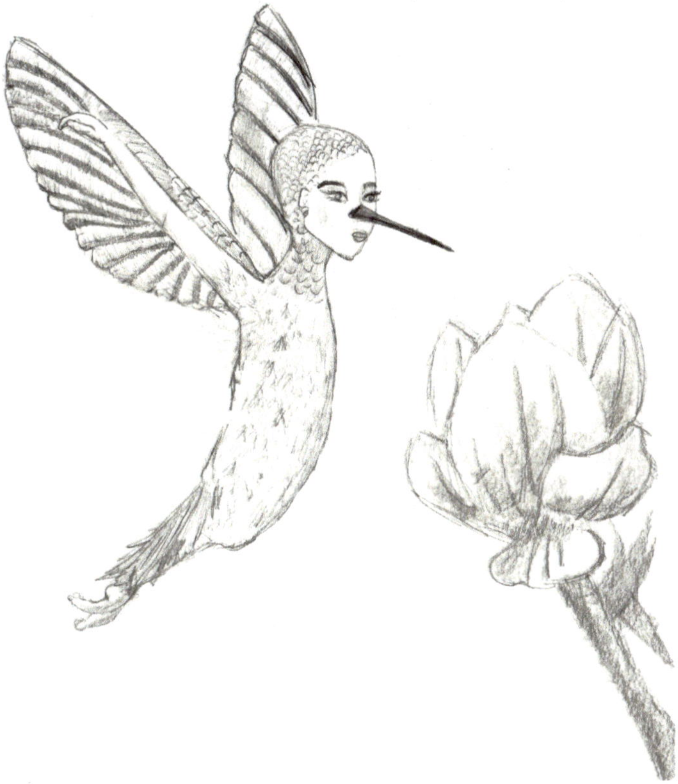

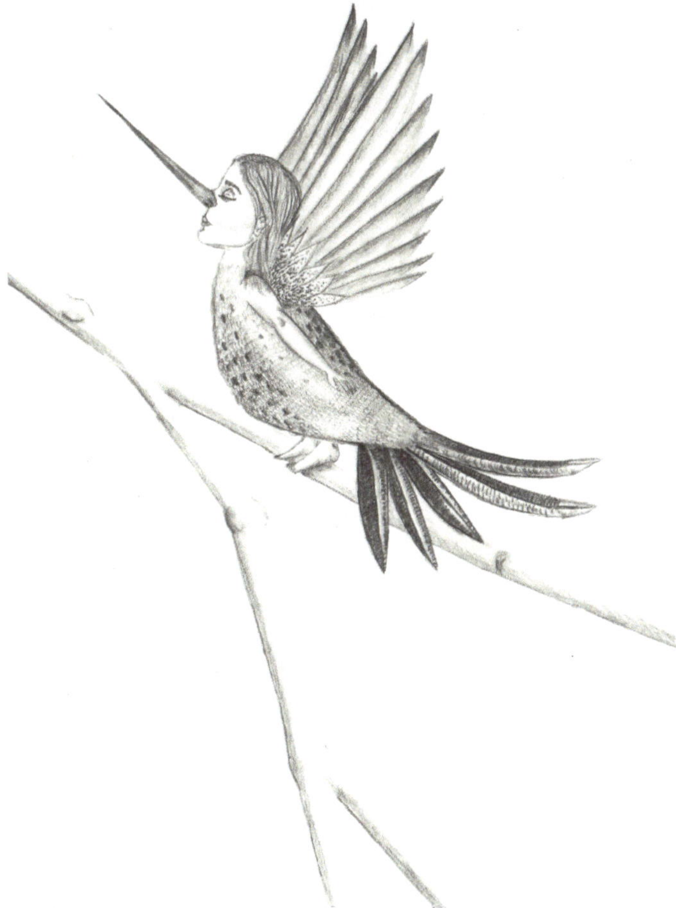

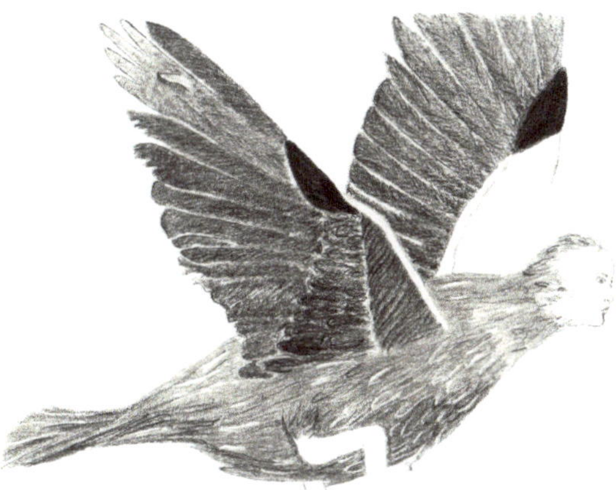

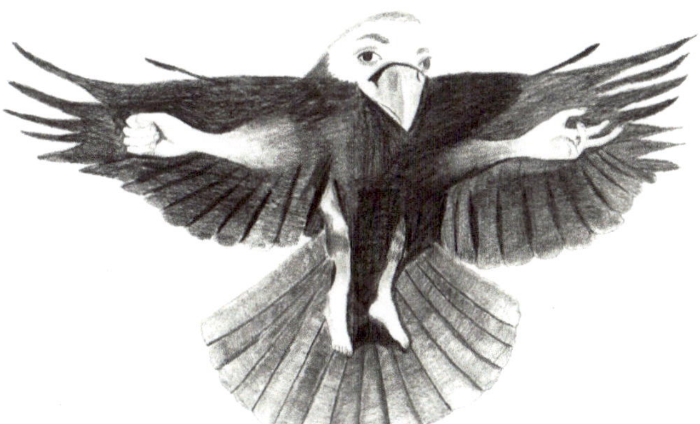

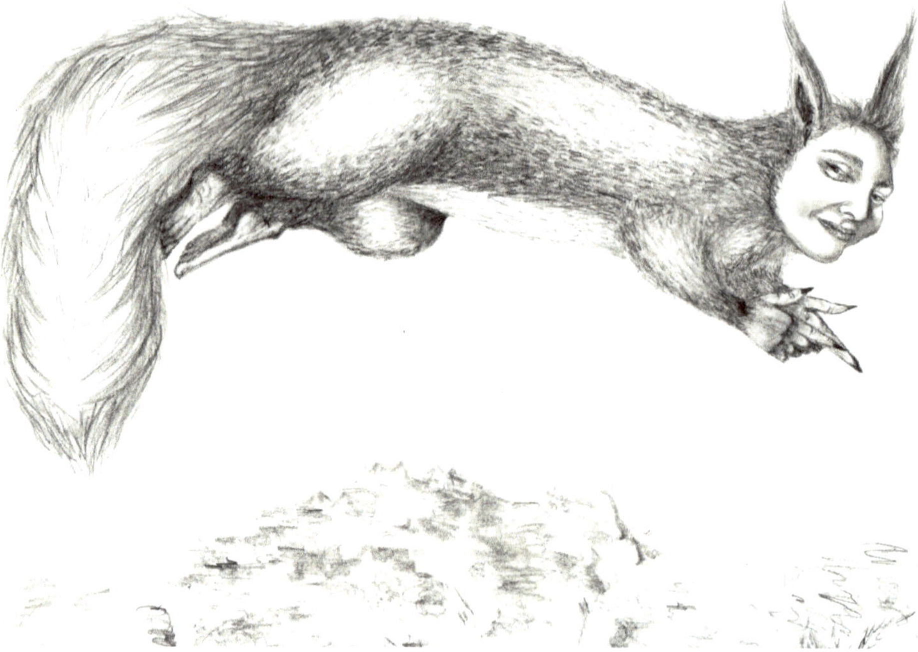

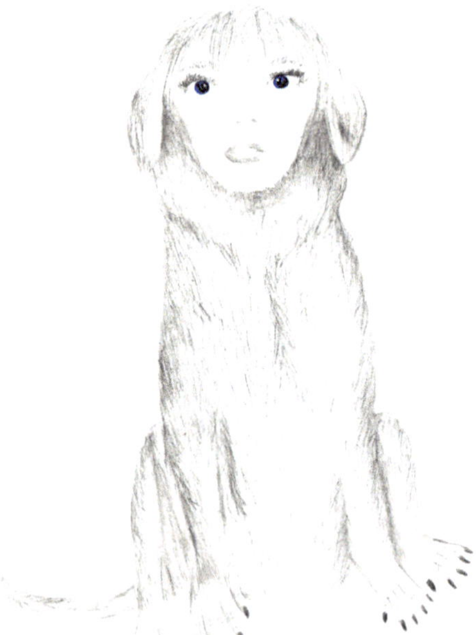

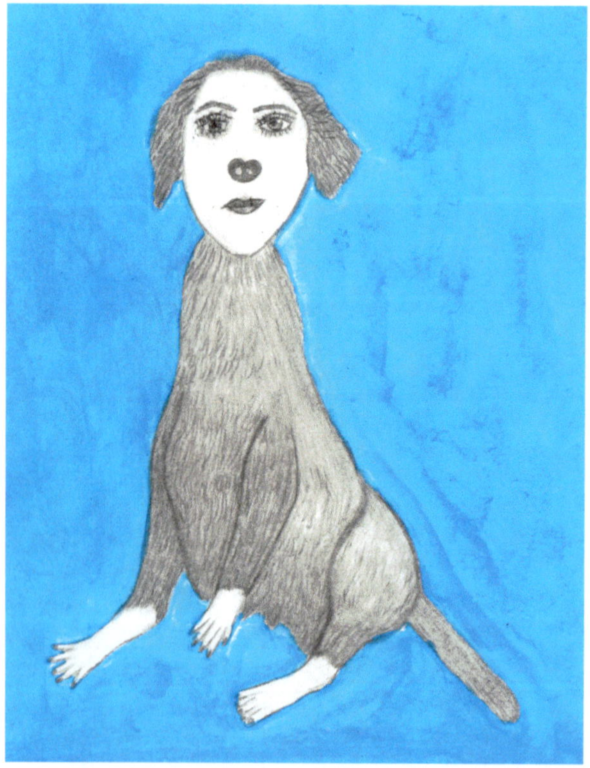

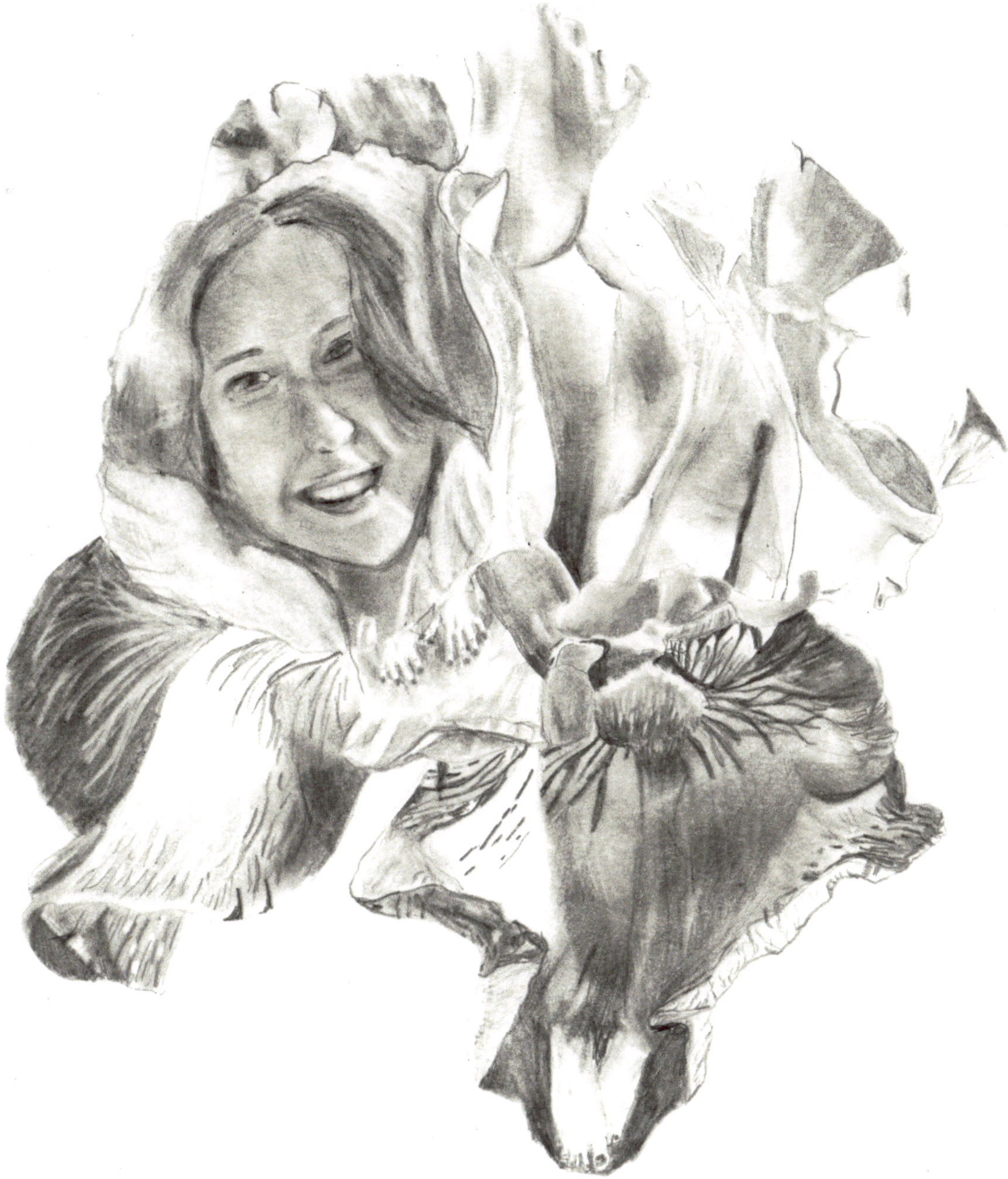

IRDS • BLACKBIRD • CANADA GEESE • CHICKEN • CROW • DOVE • D

EAGLE • EGRET • FINCH • FLICKERS • HAWK • HERON • HUMMINGI

MAGPIE • MOCKINGBIRD • OWL • SCRUB JAY • STARLING • SWALLO

ANAGER • TURKEY • TURKEY VULTURE • LANDSCAPE • CLOUDS • D

REEZE • FLATNESS • FOG • NORTH WIND • PUTAH CREEK • SUNS

LANTS • ACORN • BERMUDA GRASS • ELDERBERRY • BUCKEYE • BUL

CEANOTHUS • CENTURY PLANT • CHINESE PISTACHE • CHINESE TAL

COTTONWOOD • DAFFODIL • DATURA • DISCARDED CHRISTMAS T

ELDERBERRY • FIG • FLOWERING QUINCE • FOXTAIL GRASS • GINK

ACKBERRY • HEUCHERA • IRIS • JASMINE • LEAF PILE • LEMON • LU

MILKWEED • MISTLETOE • MODESTO ASH • MUSTARD • OLIVE TR

RNAMENTAL PEAR • PENSTEMON • PERSIMMON • PINE • POMEGRAN

POPPY • PURPLE NEEDLE GRASS • RED HOT POKER • REDBUD • REDW

ROSEMARY • SALVIA • SOURGRASS • SUNFLOWER • SYCAMORE • TOM

TOYON • TULE • VALLEY OAK • WALNUT • FISH AND REPTILES • FR

ALMON • TURTLE • MAMMALS • CAT • COW • COYOTE • DOG • HOR

TTER • OPOSSUM • RABBIT • RACCOON • RAT • SQUIRREL • INSECTS • B

LACK WIDOW SPIDER • BUTTERFLY • COCKROACH • CRANE FLY • DEE

DRAGONFLY • FLY • MOSQUITO • MOTH • PRAYING MANTIS • FANT

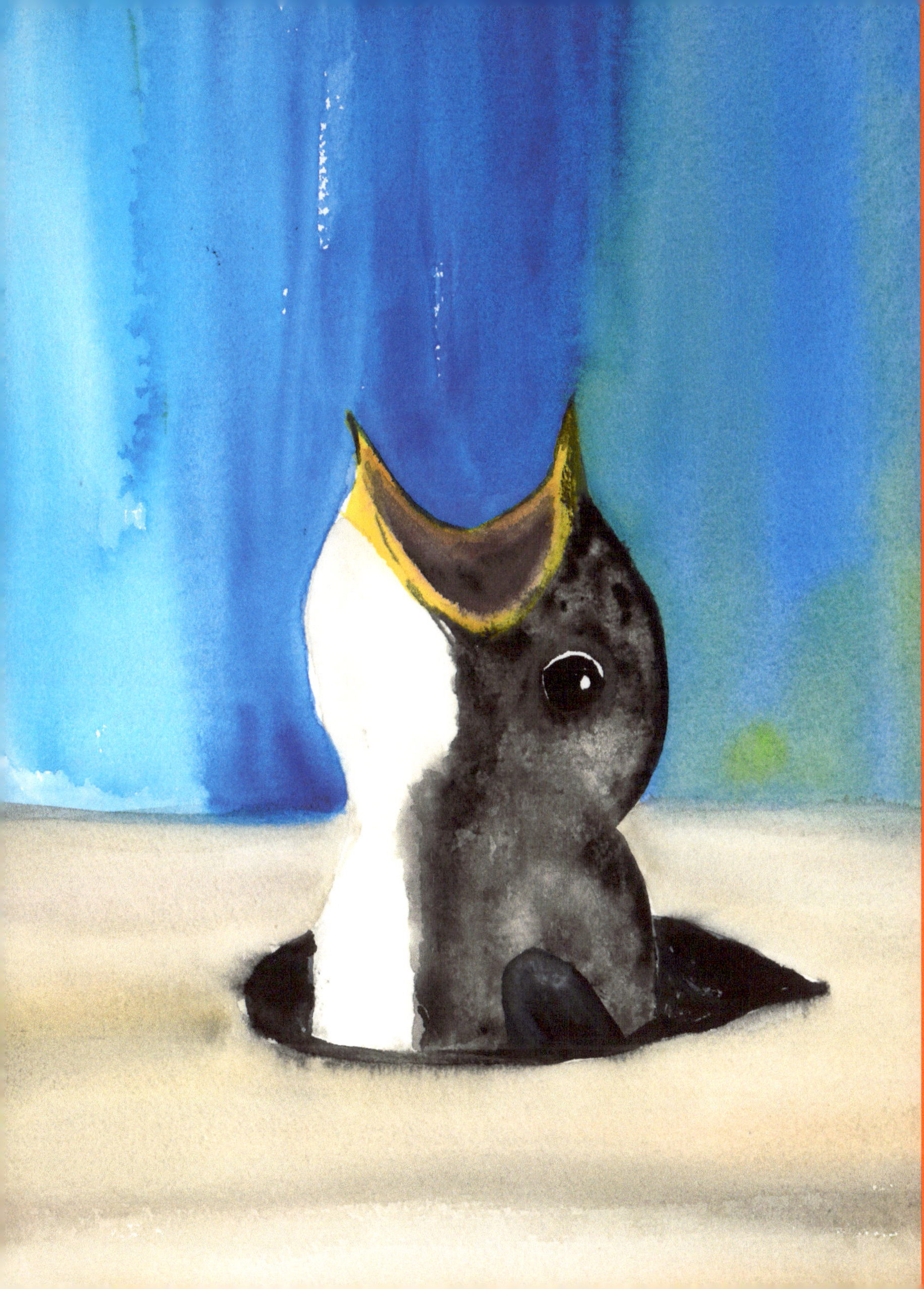

BIRDS

The following artists created the artwork in this section. The list is alphabetical by category and does not necessarily represent the order nor page number the artwork appears; please refer to the page numbers next to the category to locate where the art appears. Some teachers/schools did a class project (CP) for one category; all the artists involved in that project are listed in italics under the teacher and school name within the category. For other class projects the artist chose their category; their names appear under that category with a footnote indicating their teacher/school. If there is a number after an artist's name they provided more than one piece of art for that category.

BLACKBIRD
(pages 86-87)
Gianulias, Juliana CP[1]
CP, 5th, Mary Schembri, Judy Catambay, Art Teacher, Willett
A
Arthur
Bekele
Bishesha
Brynn
Bryson
Charlie
Daniel
Gwynn
Hannah
Holly
Jessyca
Julian
Julianna
Katelyn
Livy
Lucas
Luke
Maryam
Paul
Quirin
Signe
Soleil
Steven
Wesley
Yvonne

CANADA GEESE
(page 89)
Kirsch, Mary Anne

CHICKEN
(page 88)
Pollock, Nathaniel
Simpson, Ingra - CP[2]

CROW
(page 85)
Madison, Jamie
Soler, Ana Paula
Sublett, Carol
Wagner, Jerri

DOVE
(page 84)
Bassel , Claire
Dale, Yea
Dionisi, Matthew - CP[2]
Fritsch, Sydney - CP[1]
Kirk, Nancy
Perkins, Gabriella - CP[1]
Simpson, Jordan - CP[1]

DUCK
(page 88)
Equinoa, Olivia - CP[1]
Maloney, Lori

EAGLE
(page 81)
Borrego, Brooke - CP[1]

EGRET
(page 89)
Maloney, Lori

FINCH
(page 94)
Almazan, Caterina - CP[1]
Dhillon, Amrita - CP[1]
Fillatti, JoAnne (3)
McBride, Julie
Rice, Emily

FLICKERS
(pages 96-97)

CP, 6th, Amy George, Birch Lane
Berin
Gabriel
Gracia
Max
Mitchell
Roshan
Ryan
Zoe

HAWK
(pages 92-93)
Porterfield, Jesse
CP, 4th Greg Miller, Judy Catambay Art Teacher, Willett Elementary
Adelia
Alec
Arav
Aunish
Blake
Caitlyn
Connor
Emma
Finley
Grace
Henry
Jace
Jadyn
Jasmin
Jaswart
Jeffory
Jennifer
Jiwoo
Katie
Michael
Nicole
Rohan
Rowan
Sabet
Wenji

HERON
(page 95)
Jefferies, Mya - CP[1]
Maloney, Lori

HUMMINGBIRD
(pages 82-83)
Bauman, Gabby
Billena, Emily (2)
Deak, Kiera
Fitzgerald, Katelyn
Garcia, Eli
Olario, Kate
Schraner, Delia
Silva, Haley
Underhill, Carolyn

MAGPIE
(page 81)
Hartsough, Rachel

MOCKINGBIRD
(page 95)
Fleming-Mullirs, Mia - CP[1]
Kiyama, Amy - CP[1]
Malek, Hannah - CP[1]
Rhodes, Pam

OWL
(pages 90-91)
Ceniseroz, Carrie
Dahdouh, Julionah
Duran, Christian - CP[2]
Hunley, Hannah
Mariani, Brooklyn - CP[2]
Pollock, Charlie
Rodriguez, Samantha - CP[2]
CP, Melanie Bown, St. Francis High School
Berejikian, Anik
Borrego, Brooke
Frisch, Gabriella
Garaygay, Sofia
Moral, Jesiree
Pascual, Katie
Reyes, Audrey
Semon, Allison
Sevey, Shalyn

SCRUB JAY
(pages 98-99)
CP, 6th, Marla Cook, Judy Catambay Art Teacher, Willett Elementary
Adam
Alex
Andrew
Charlotte
Conor
Eliam
Emily
Esha
Hanni
Hinae
Holt
Isabelle
Joanne
Joe
Kamu
Kathy
Leyla
Liam
Marcos
Marley
Matthew
Max
Mayah
Milam
Ms. Cook
Ryan
Sam
Zachary

STARLING
(page 83)
Freeworth, Laura
Maloney, Lori

SWALLOW
(pages 79 and 83)
Maloney, Lori
Moral, Jesiree - CP[1]

TANAGER
(page 89)
Green, Lonnie
Mudd, Sarah - CP[1]
Wagner, Madeline

TURKEY
(page 81)
Maloney, Lori

TURKEY VULTURE
(page 81)
Maloney, Lori

[1]Student's Choice Class Project, Teacher: Melanie Bown, St. Francis High School • [2]Student's Choice Class Project, Teacher: Carrie Ceniseroz, St. James School

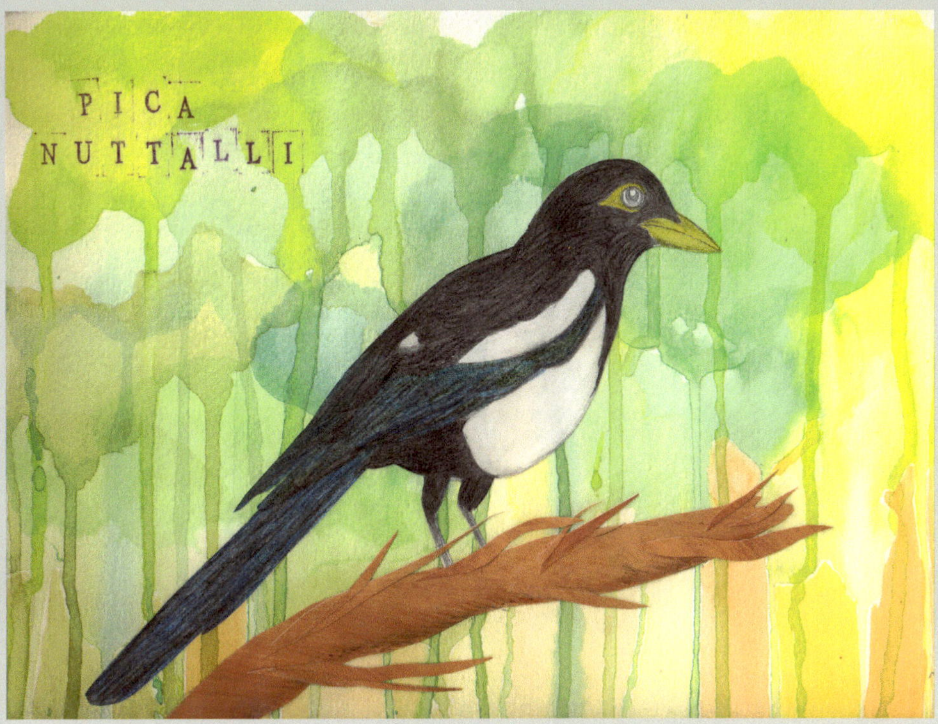
PICA
NUTTALLI

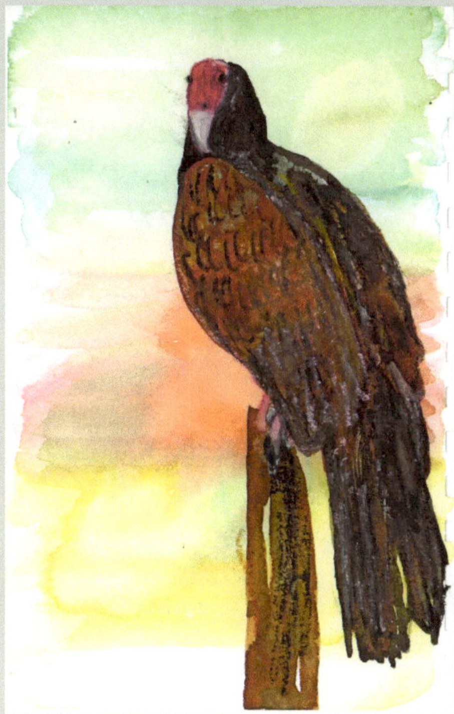

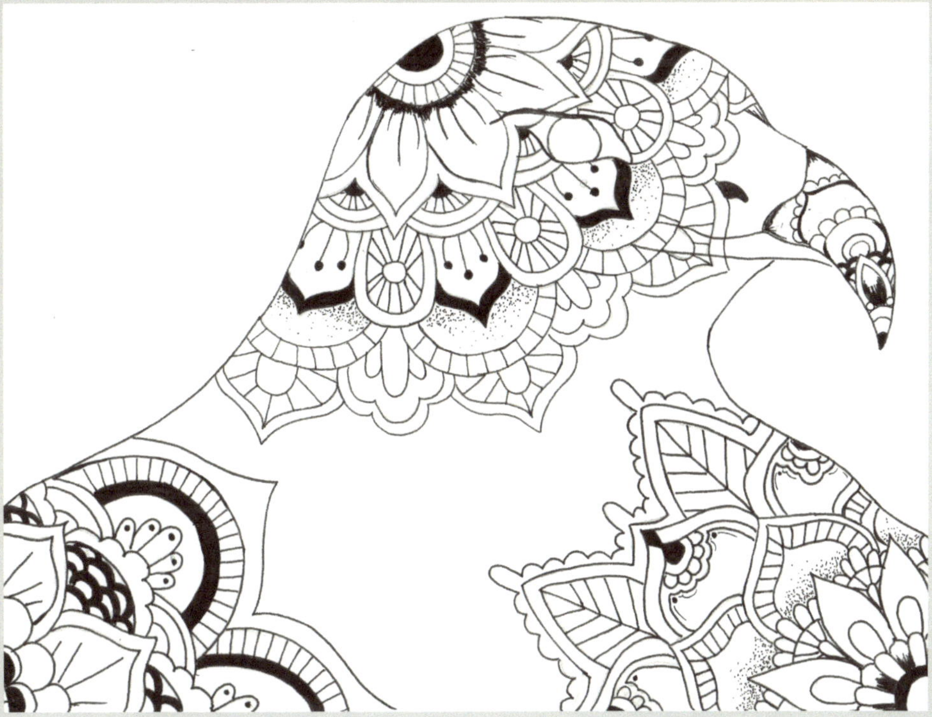

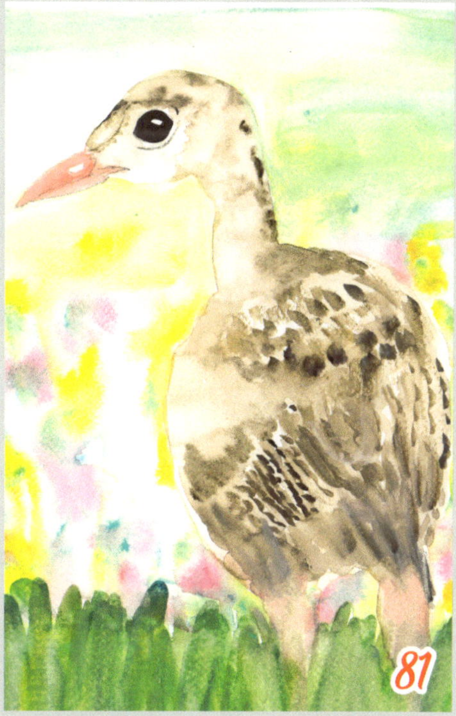

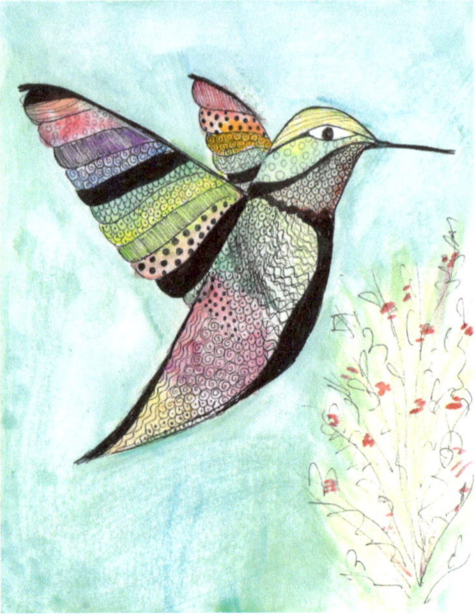
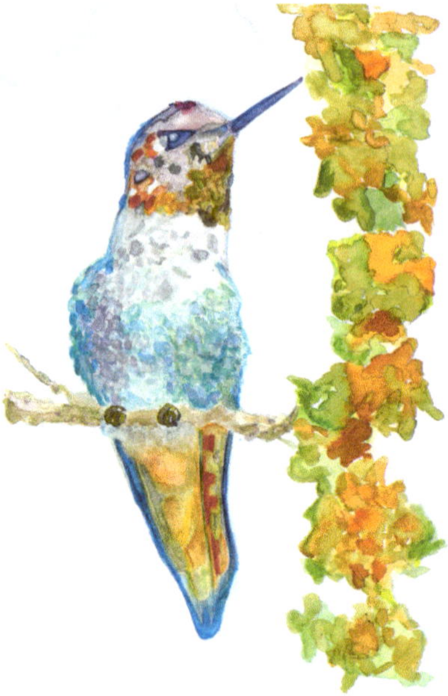
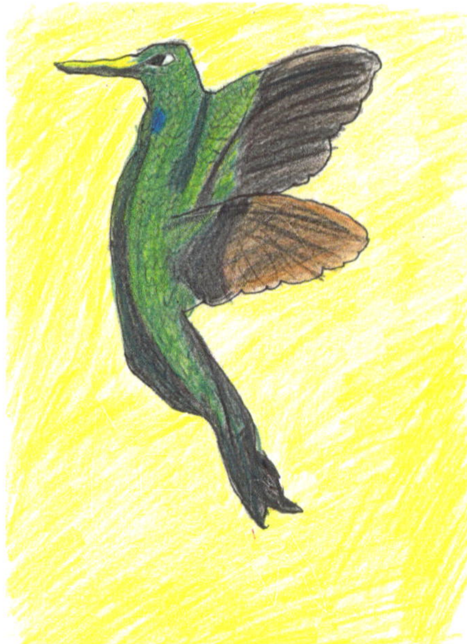
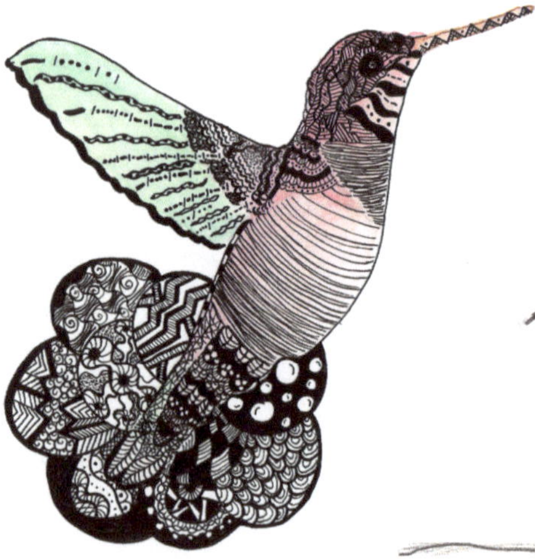
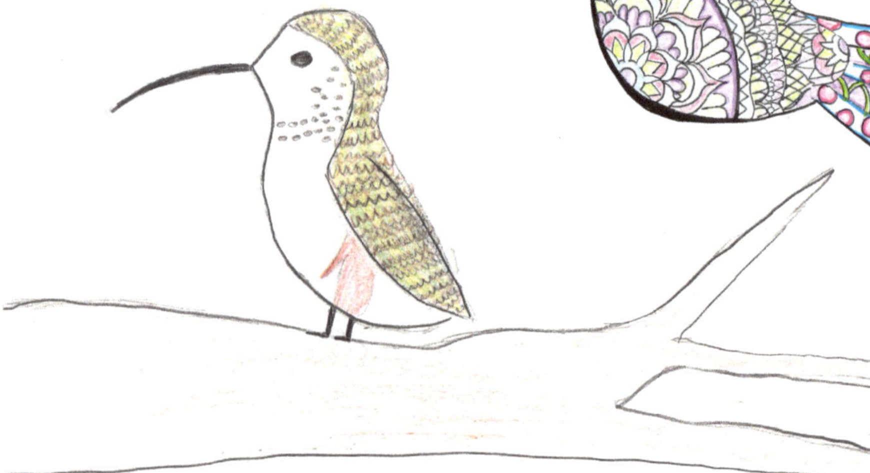
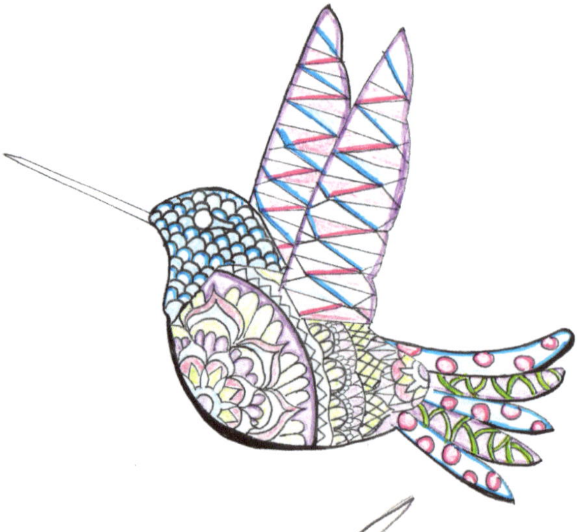

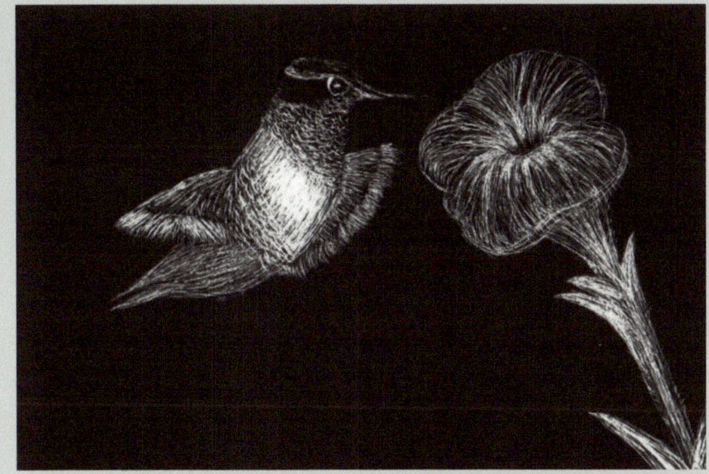

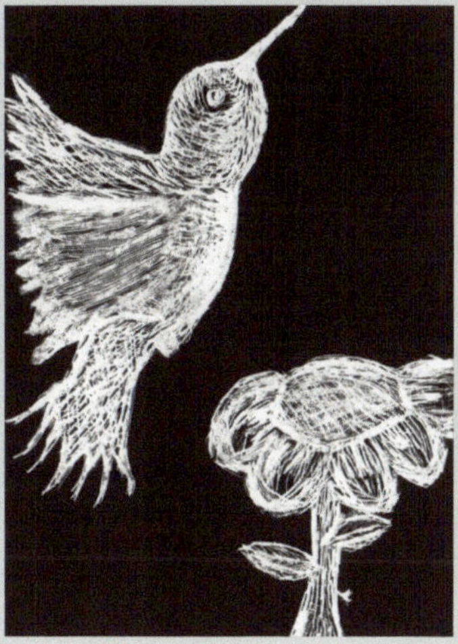

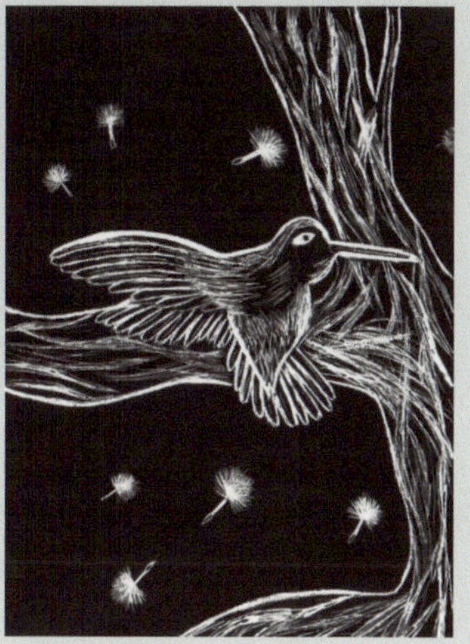

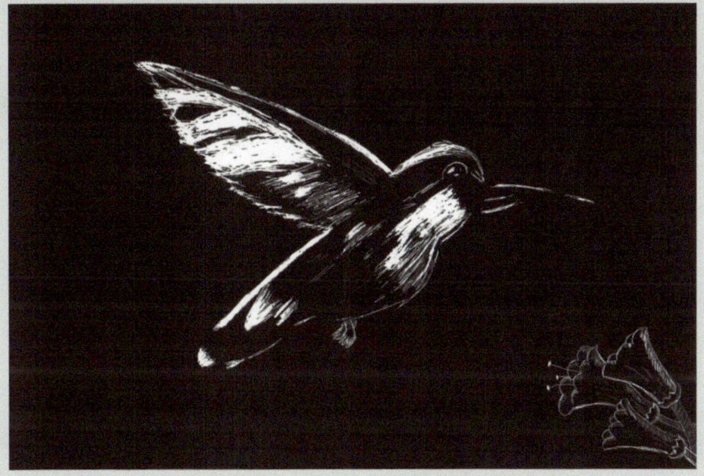

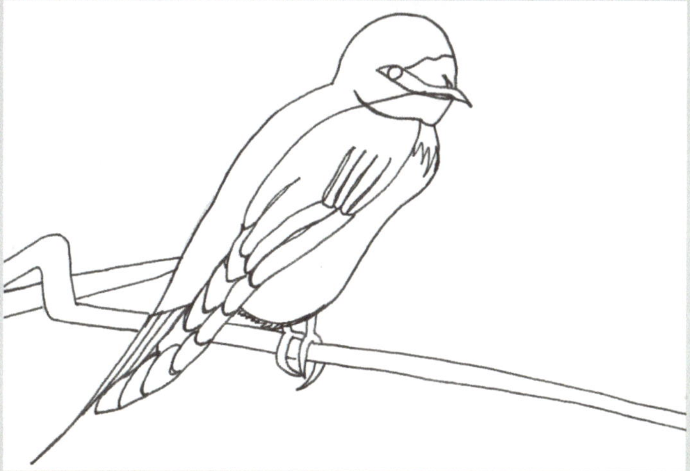

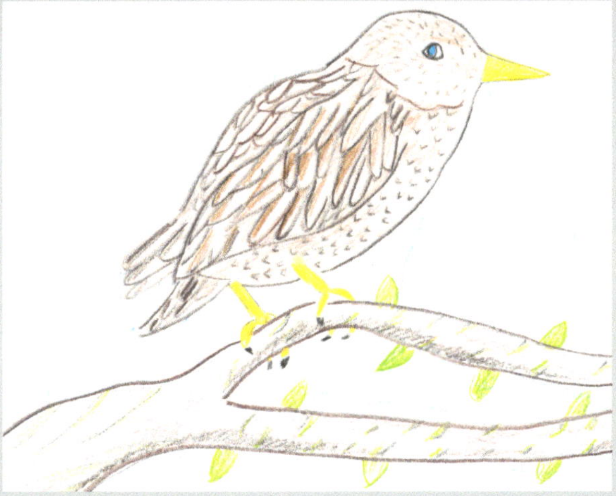

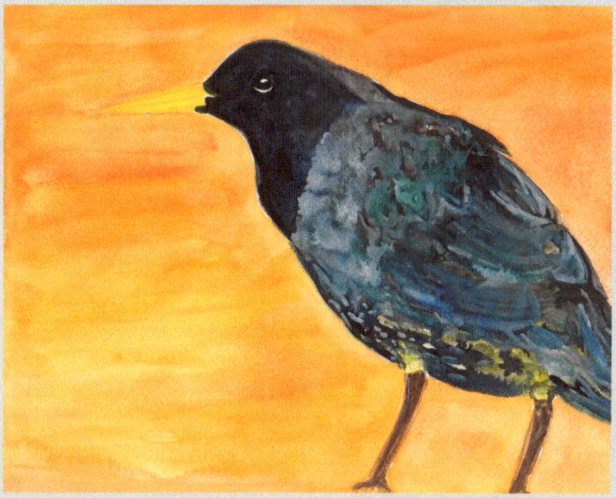

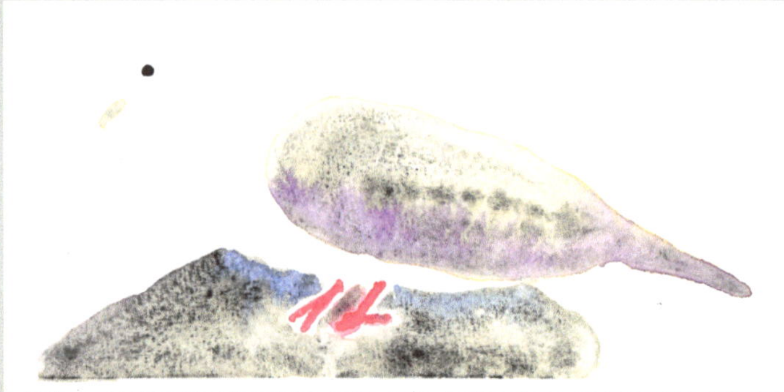

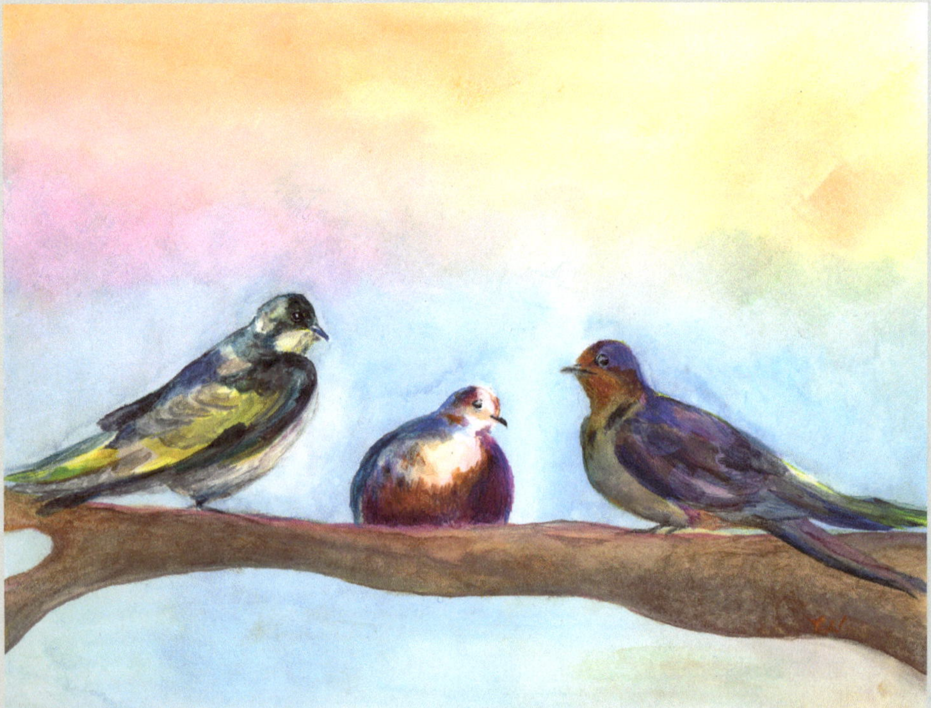

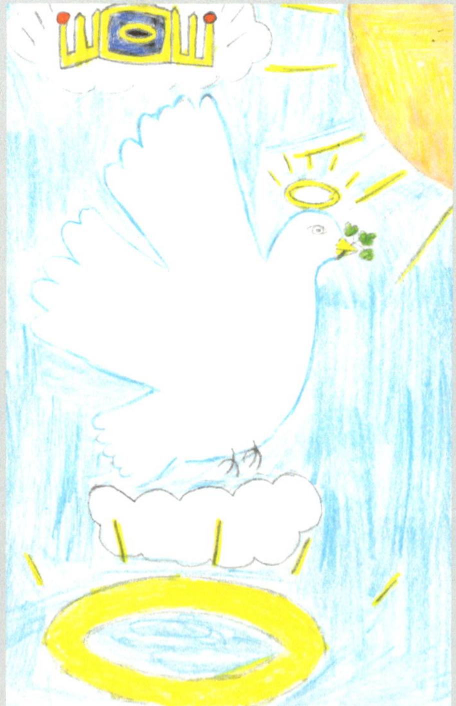

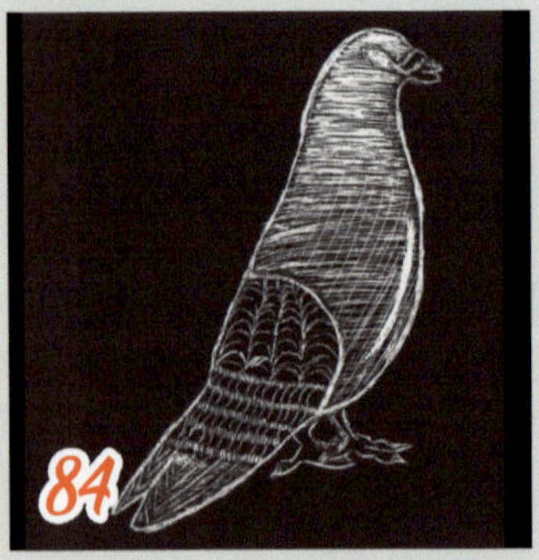

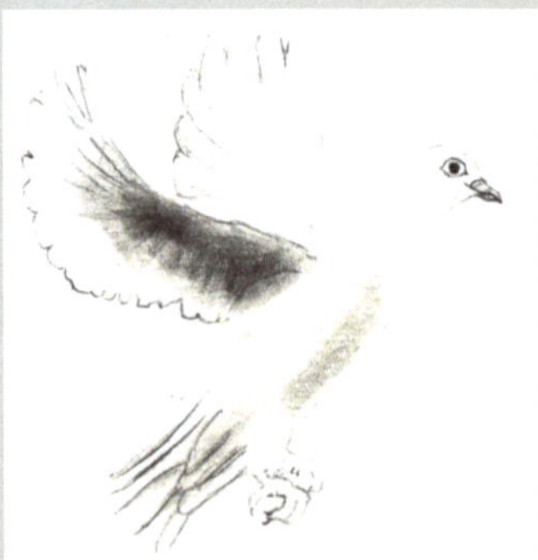

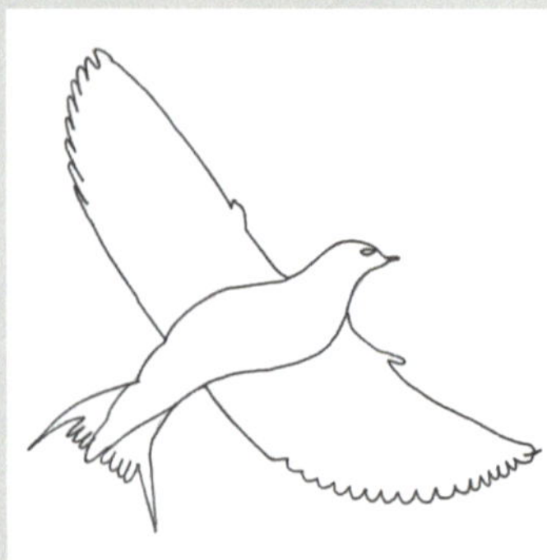

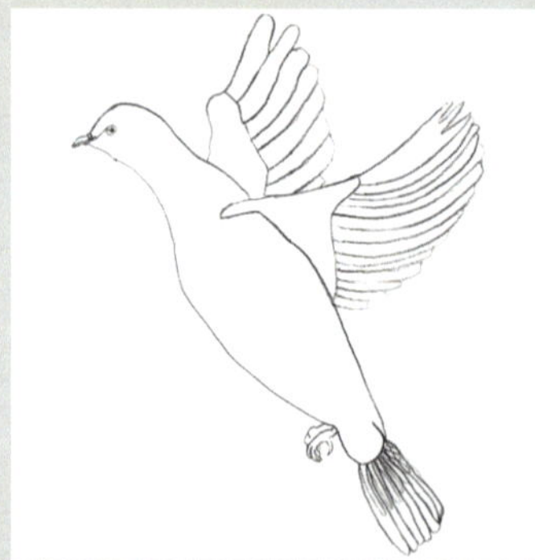

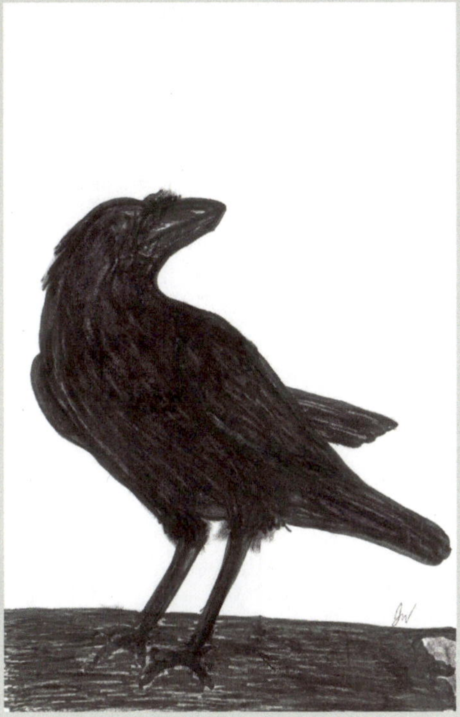

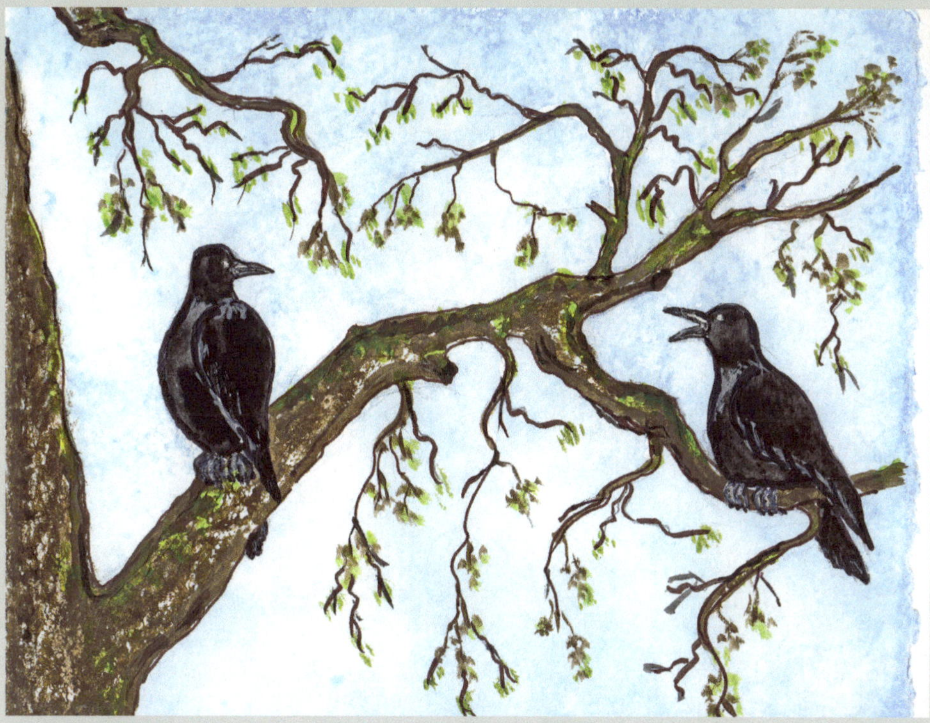

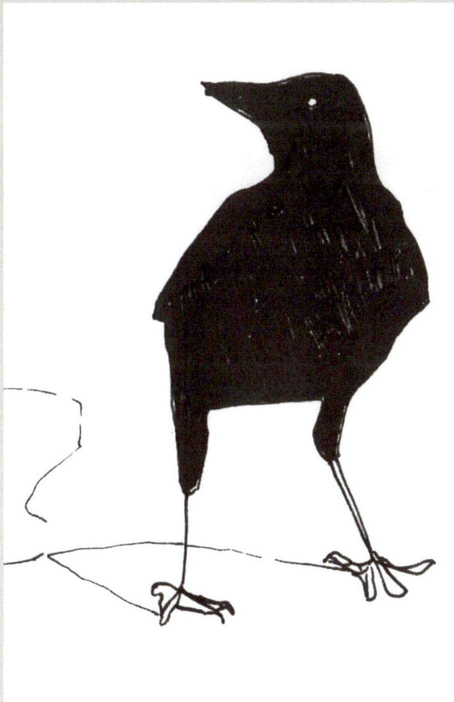

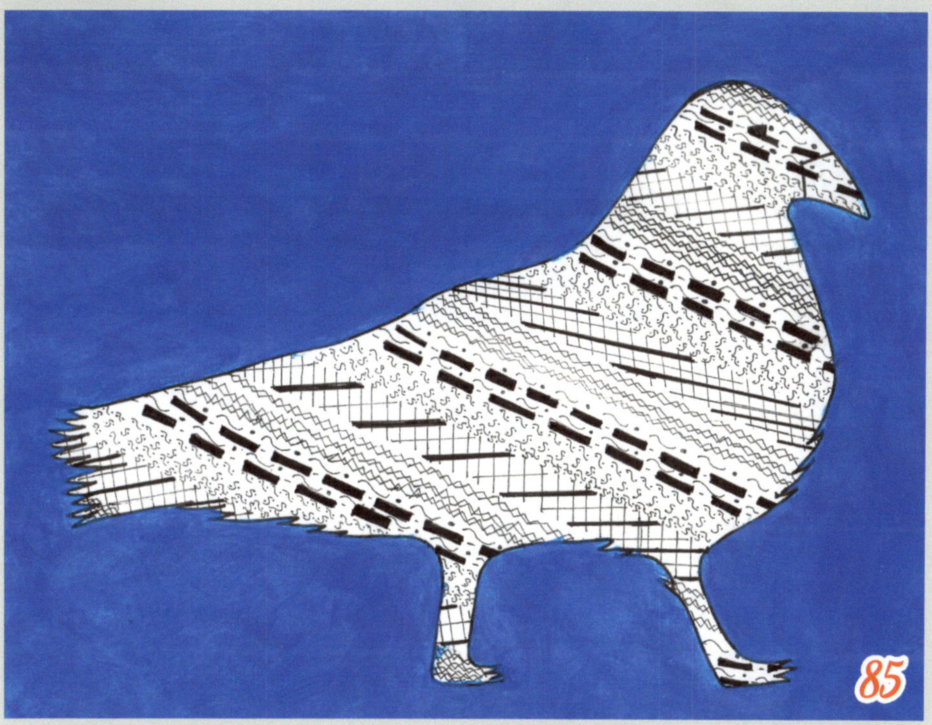

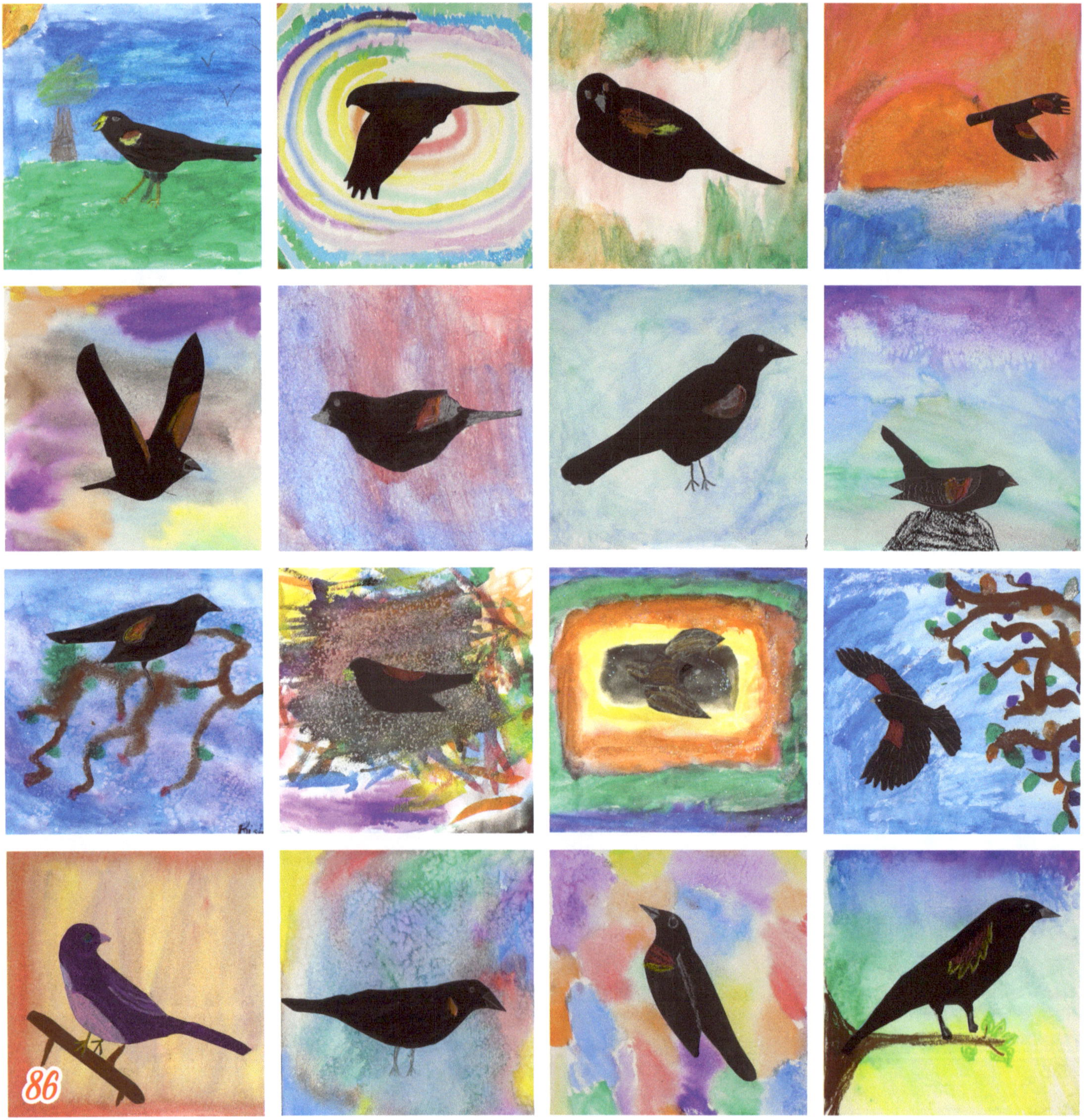

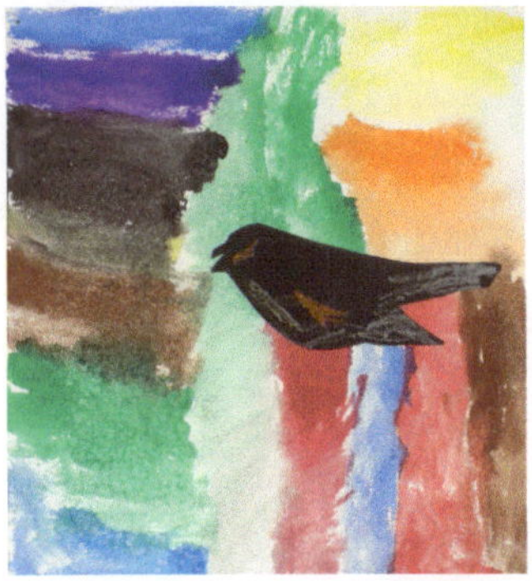
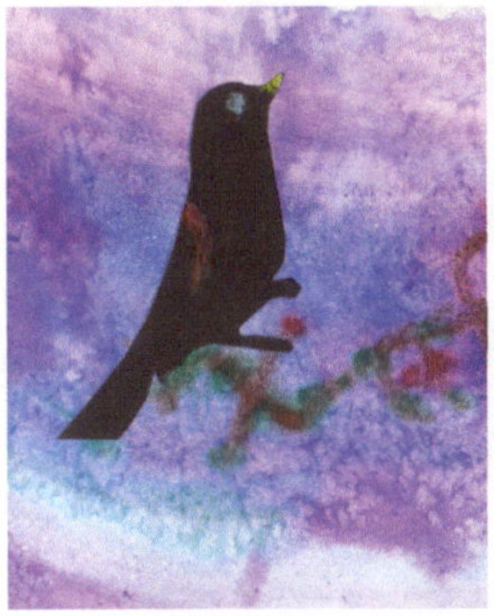
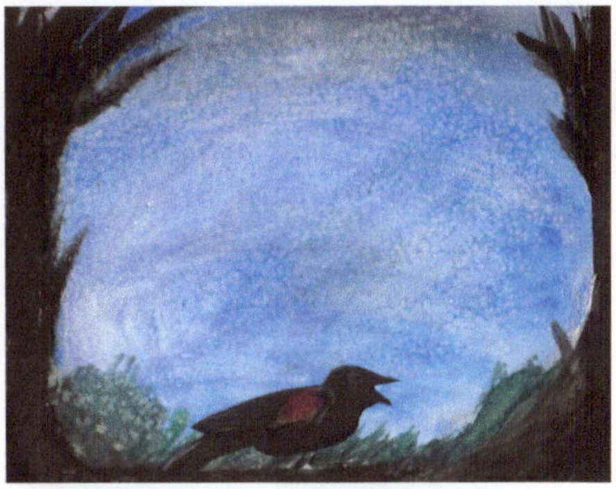
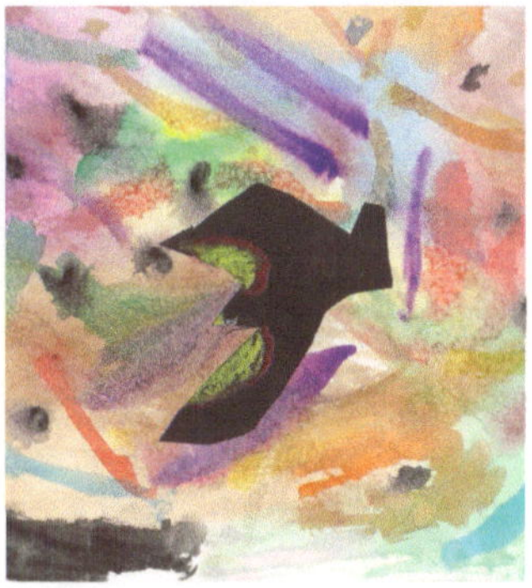
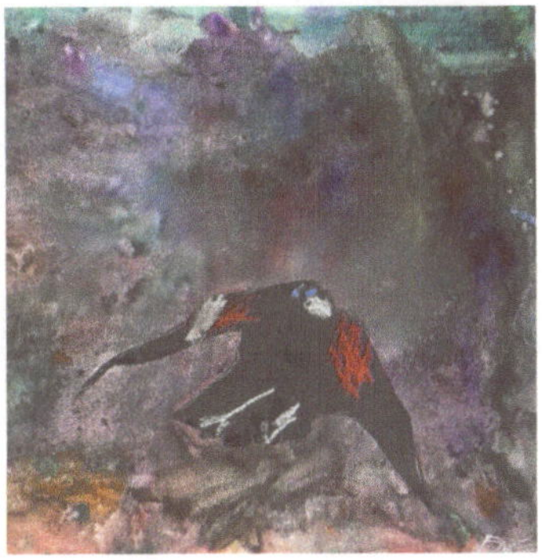
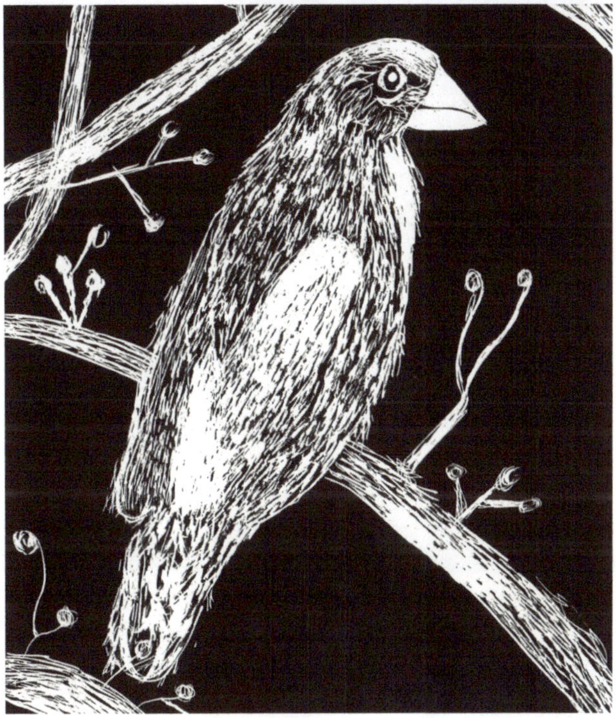
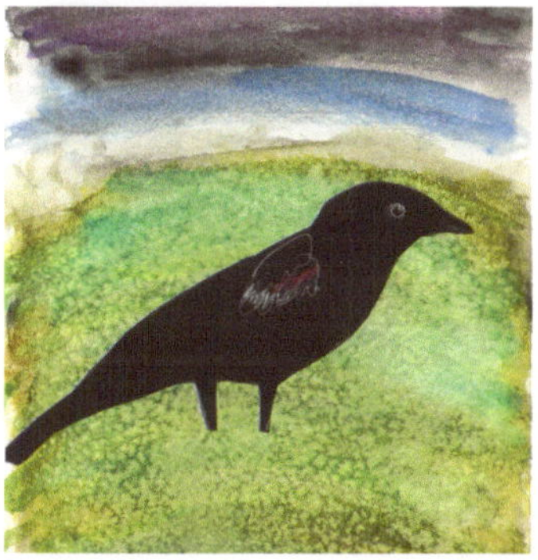
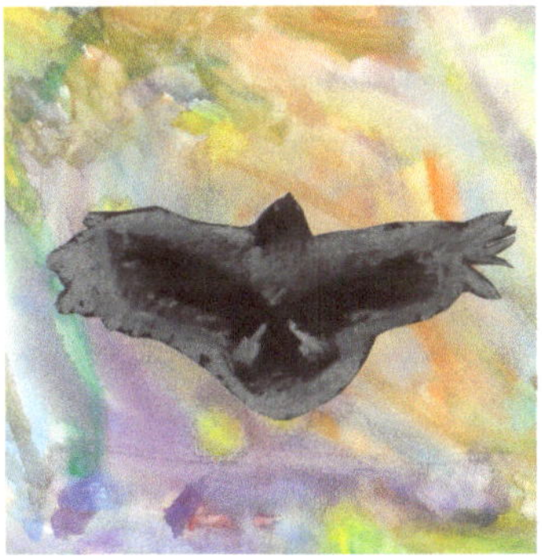
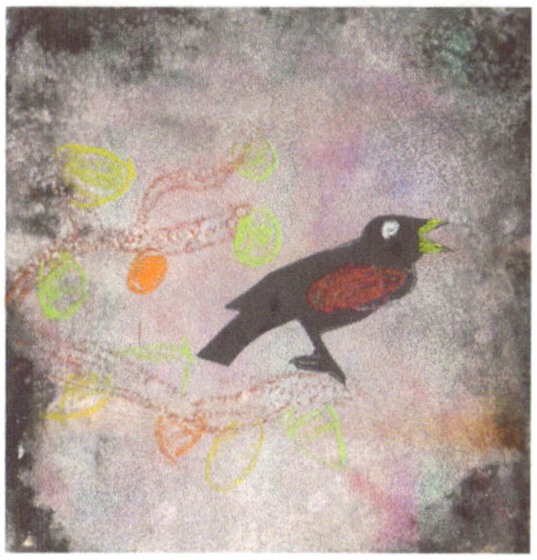
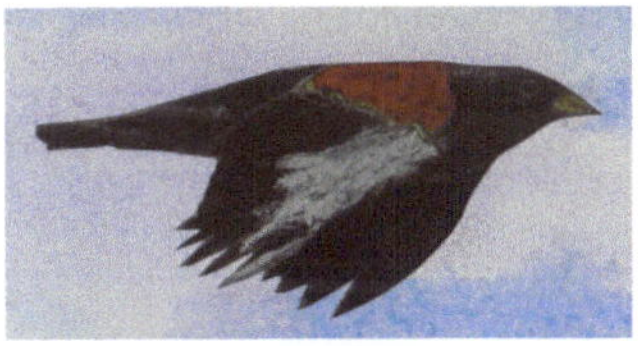
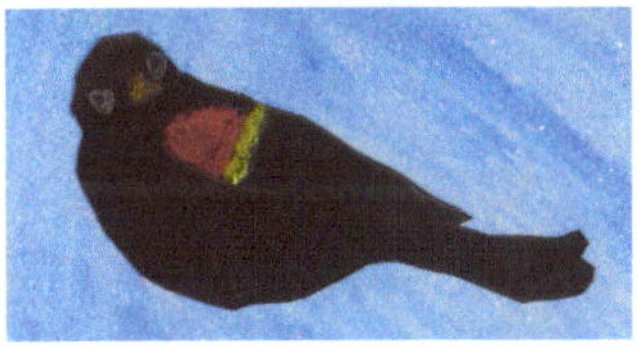
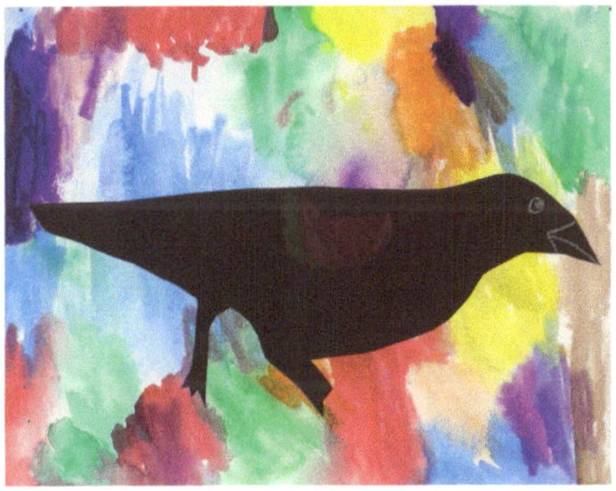
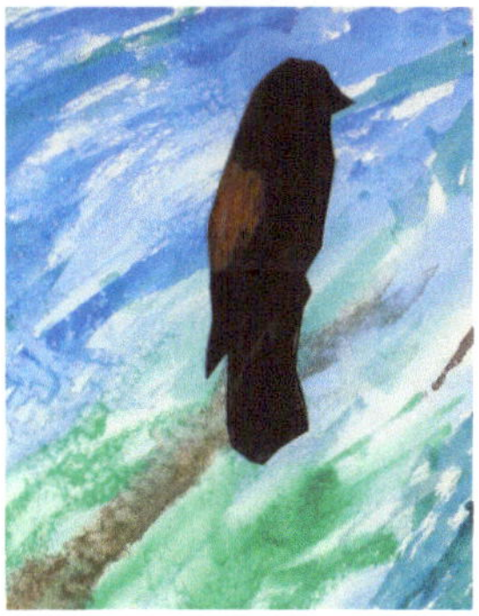
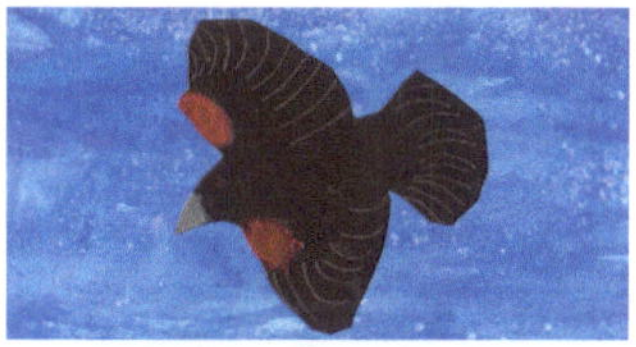
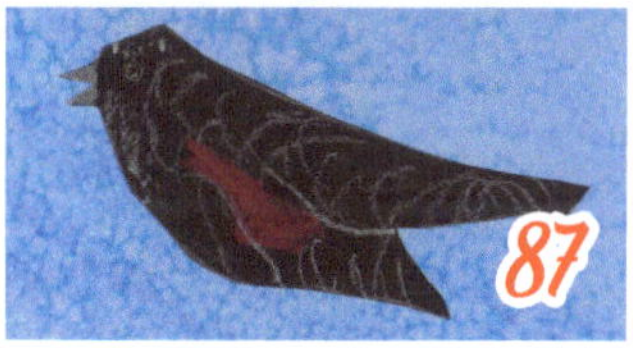

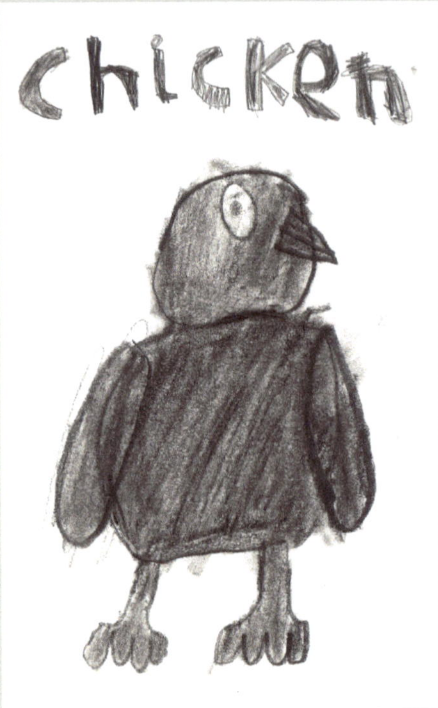
Chicken
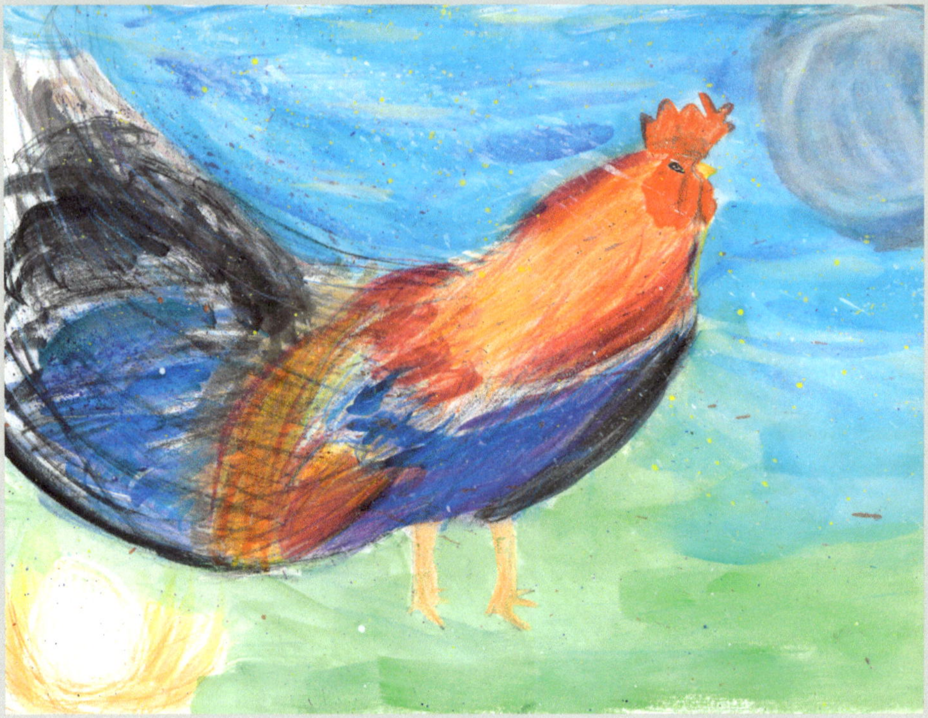
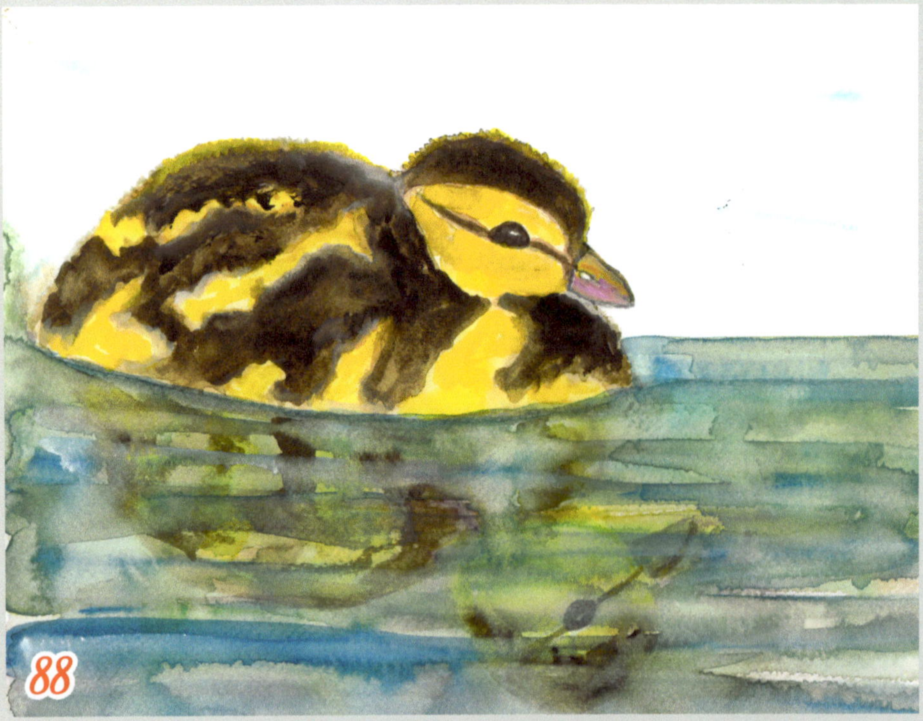
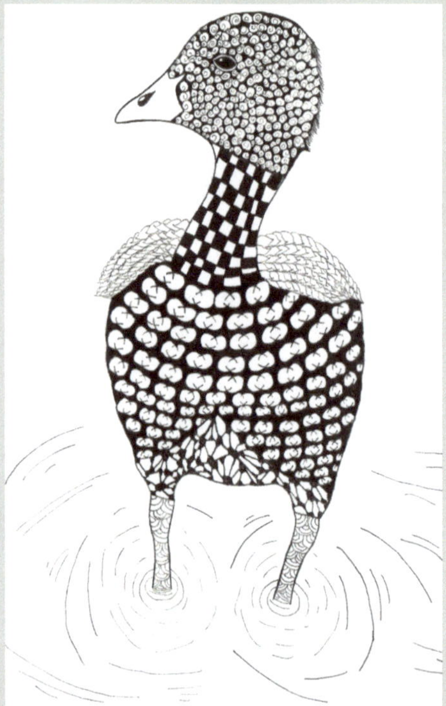

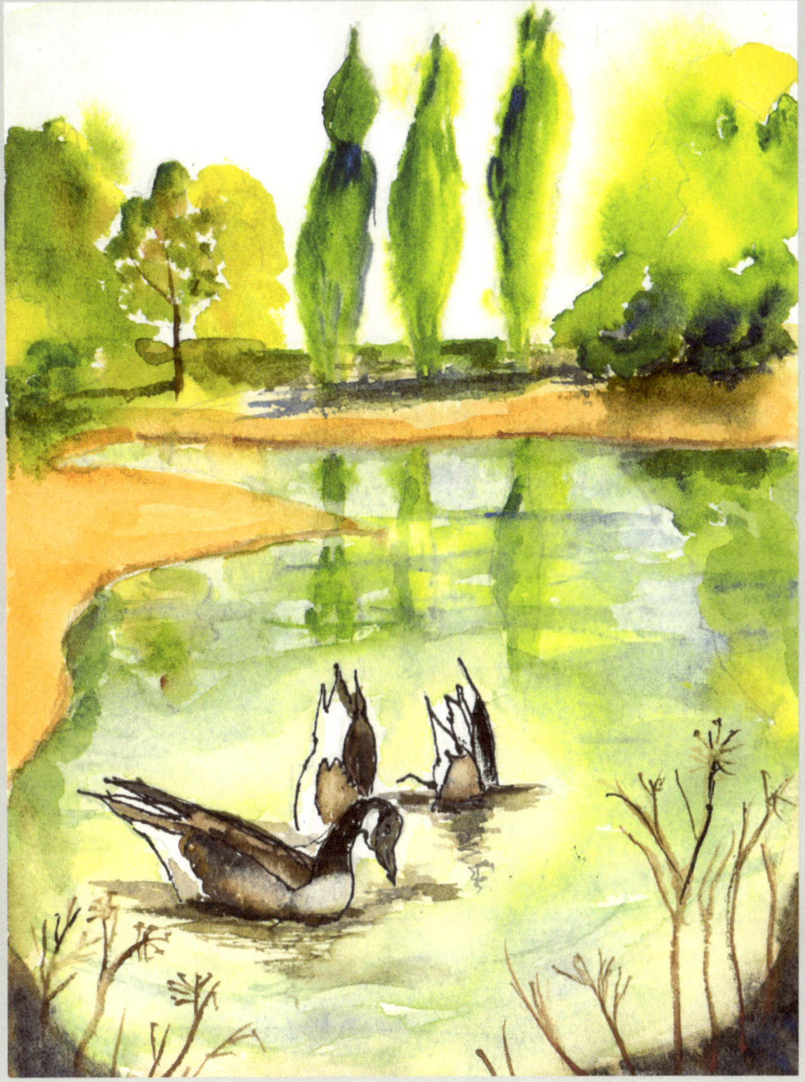
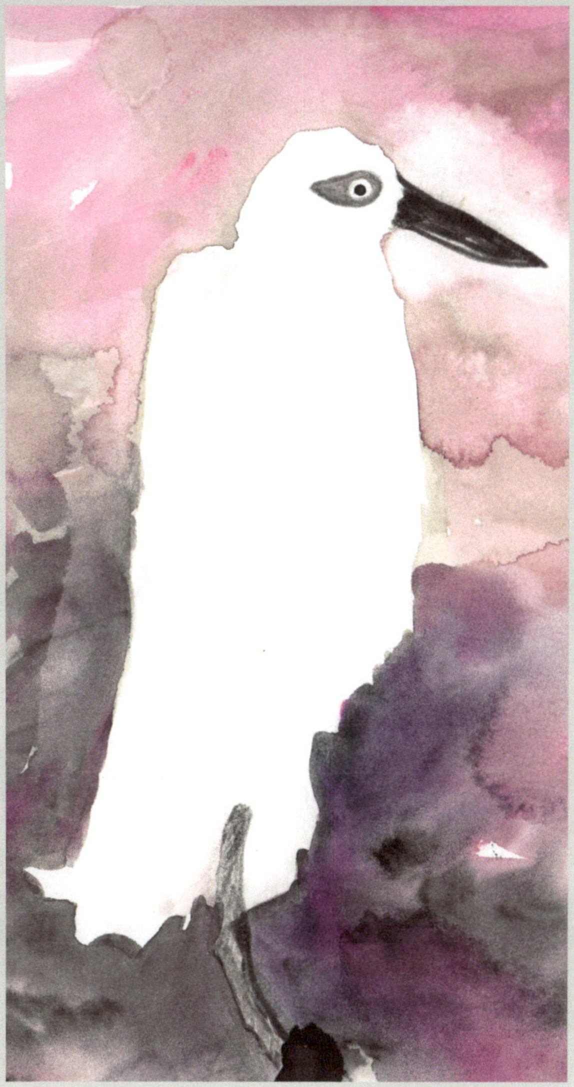
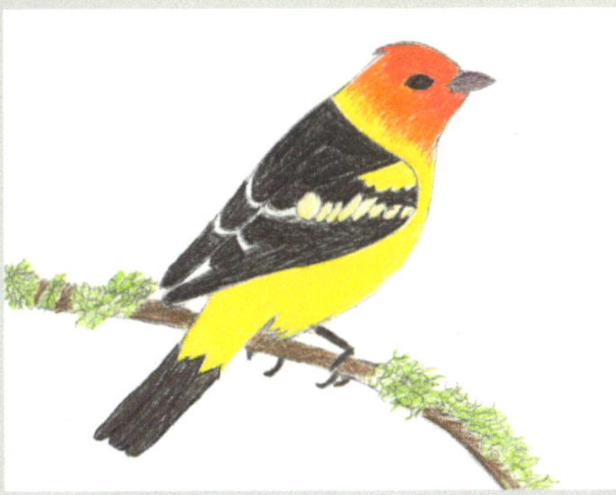
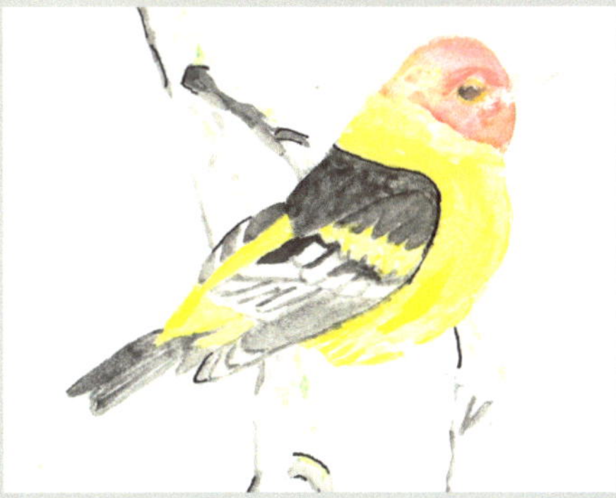
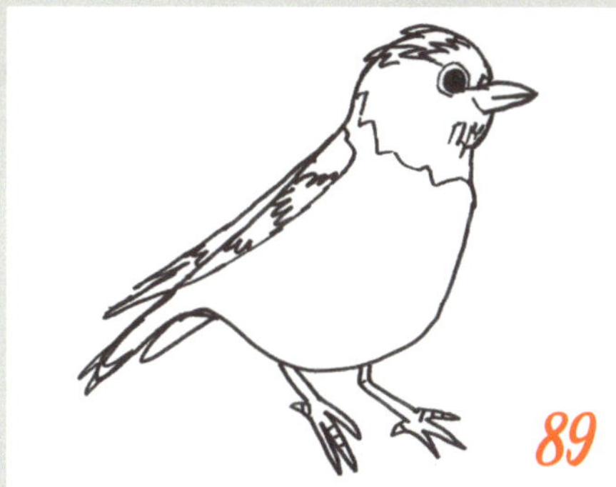

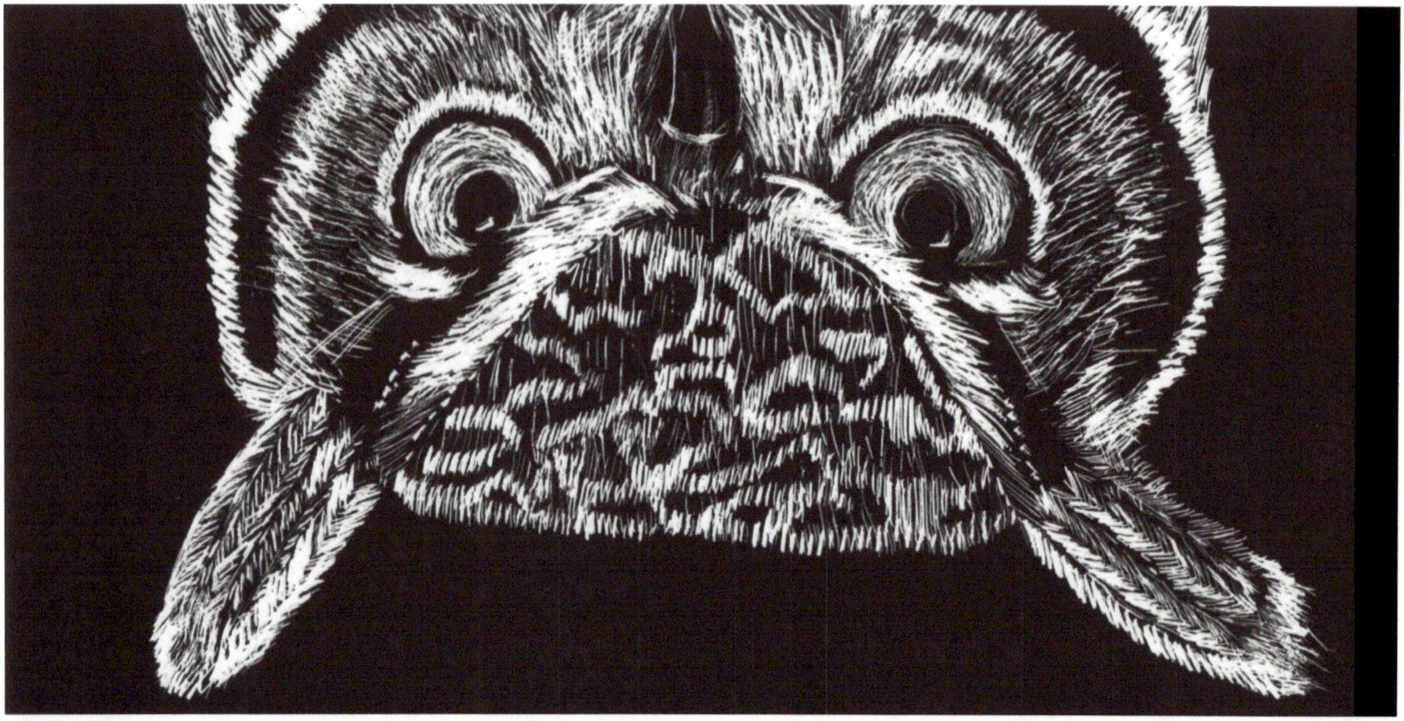

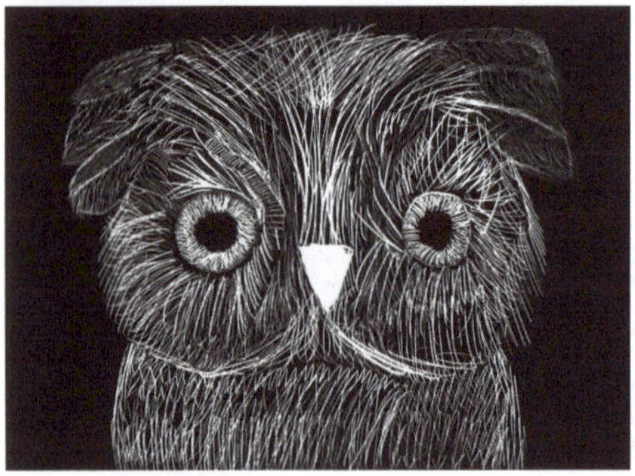

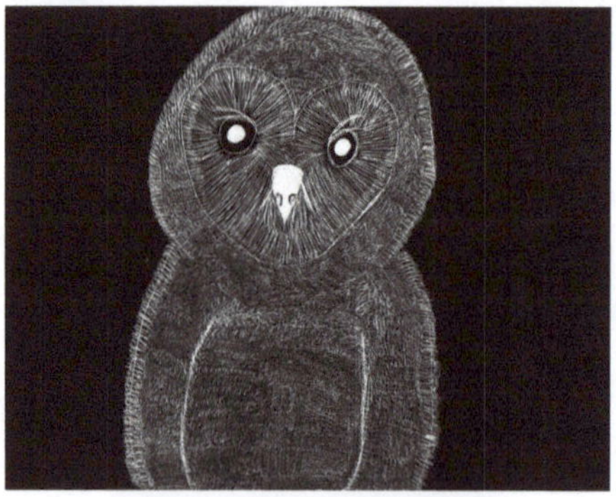

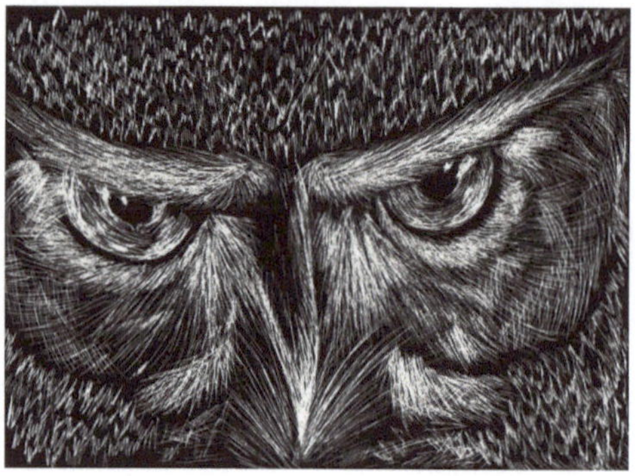

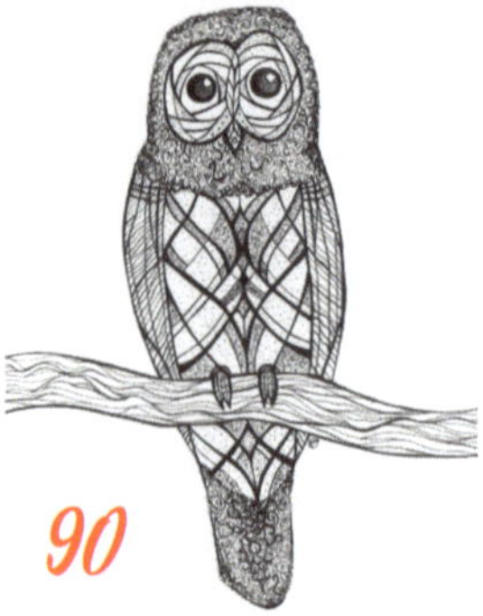

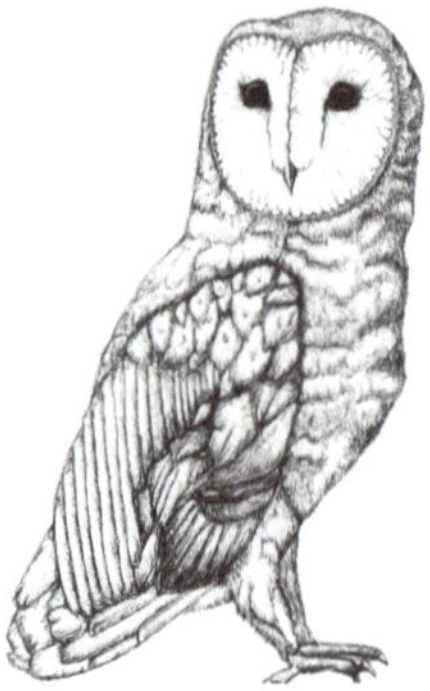

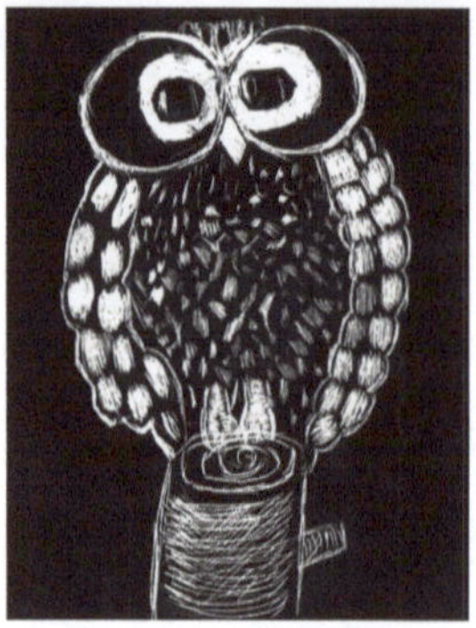

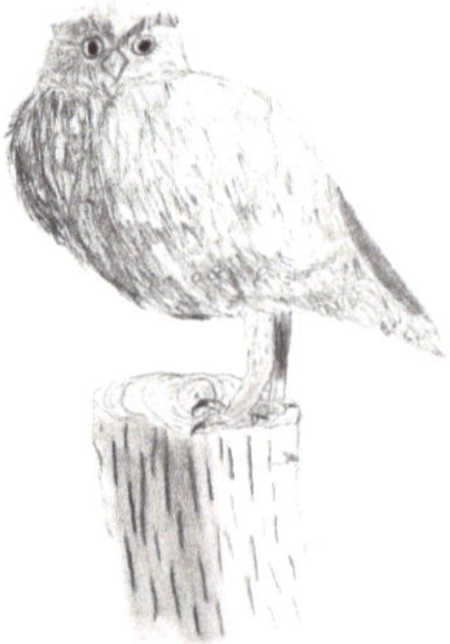

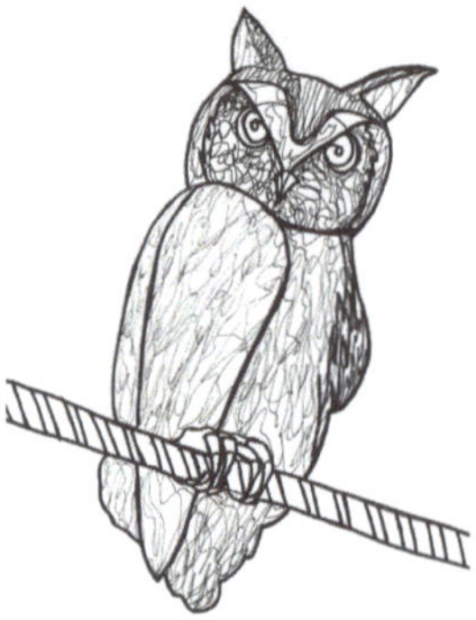

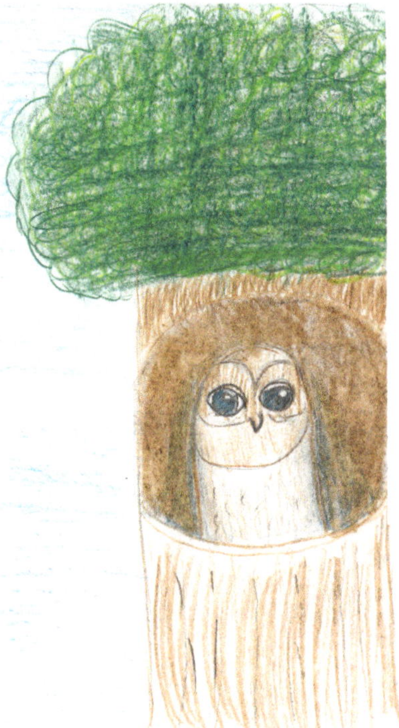

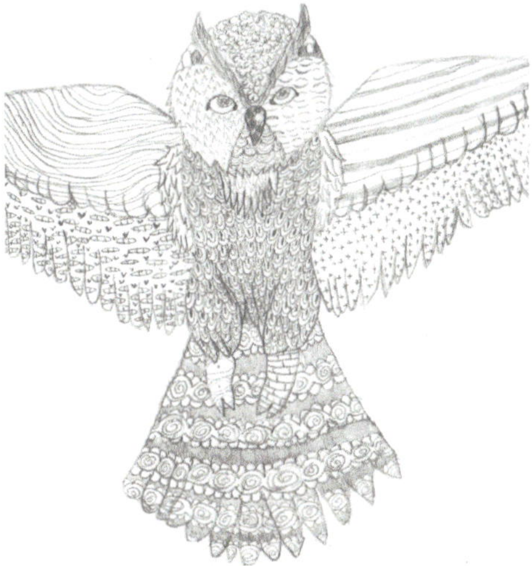

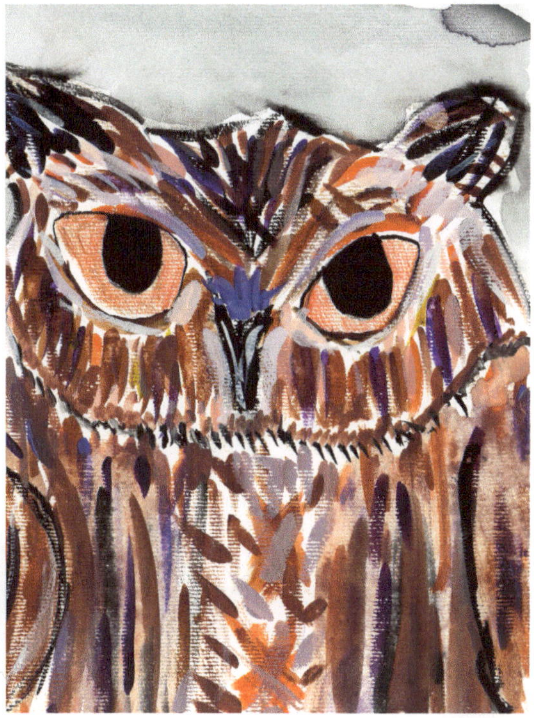

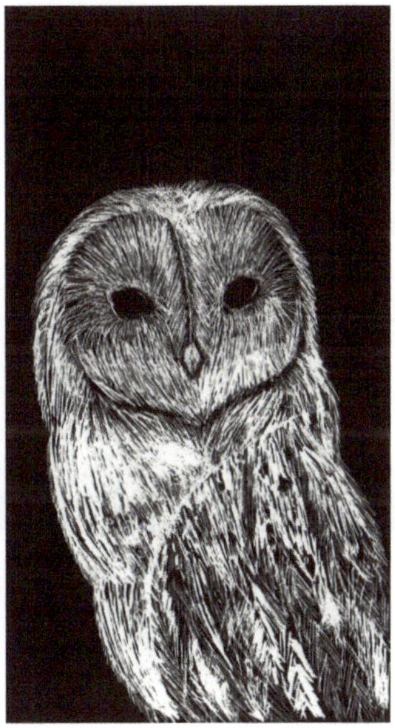

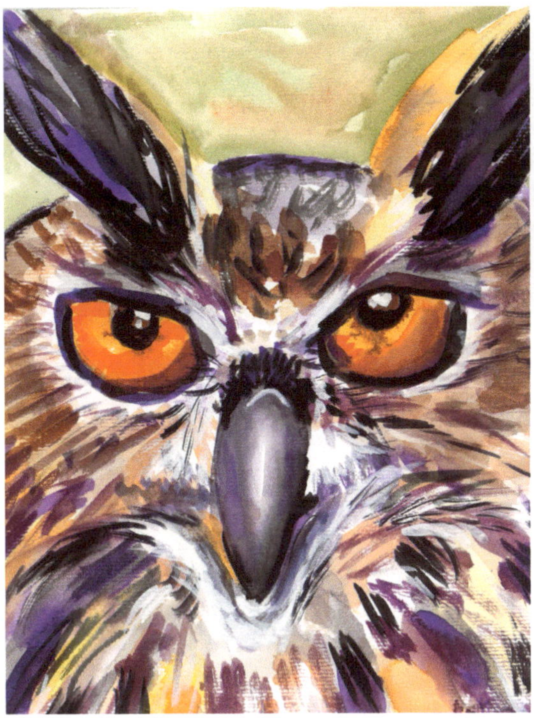

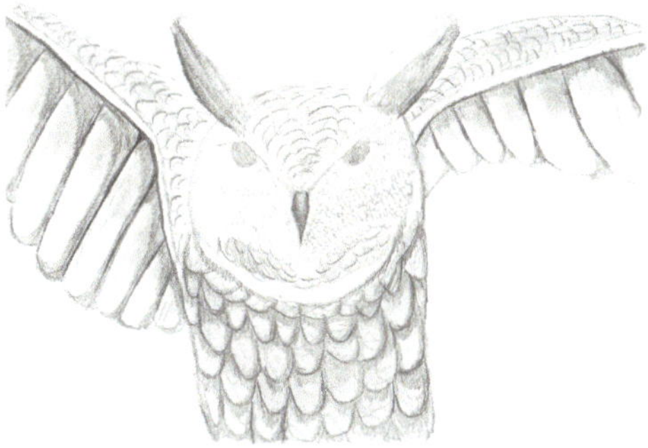

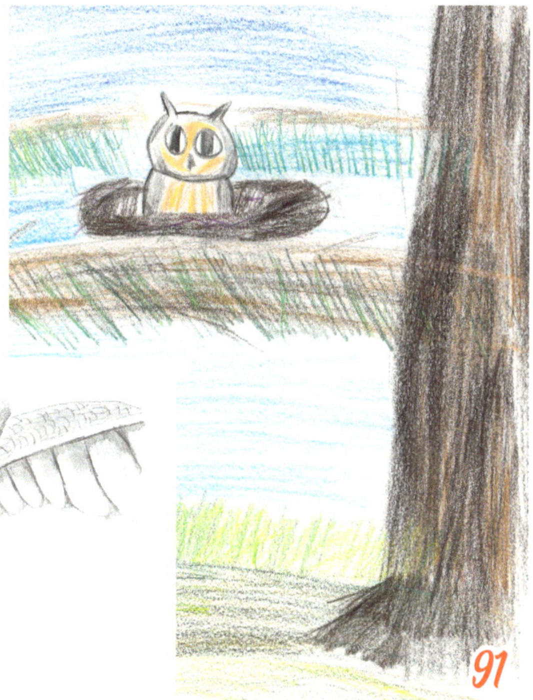

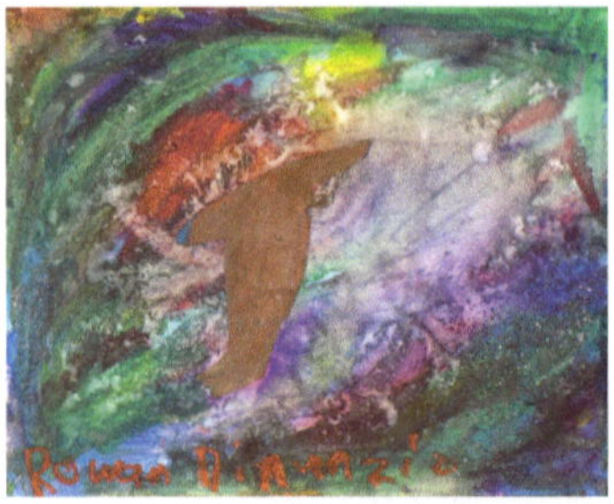
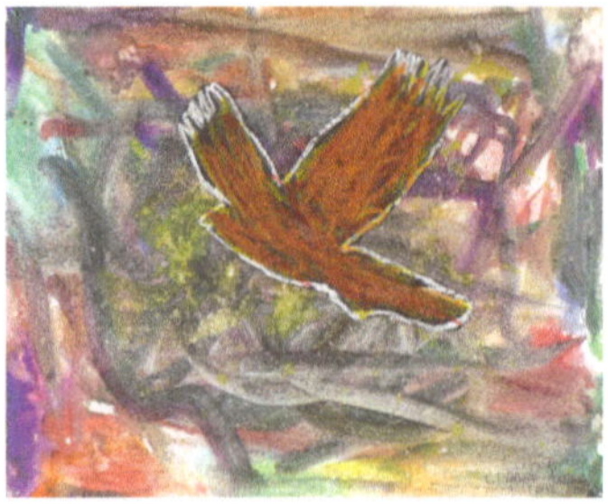
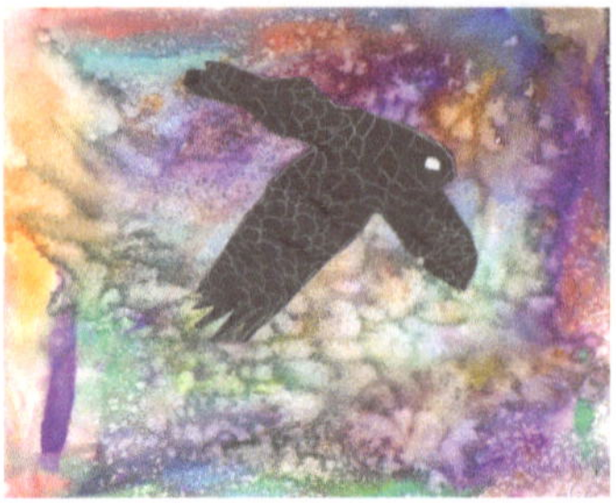
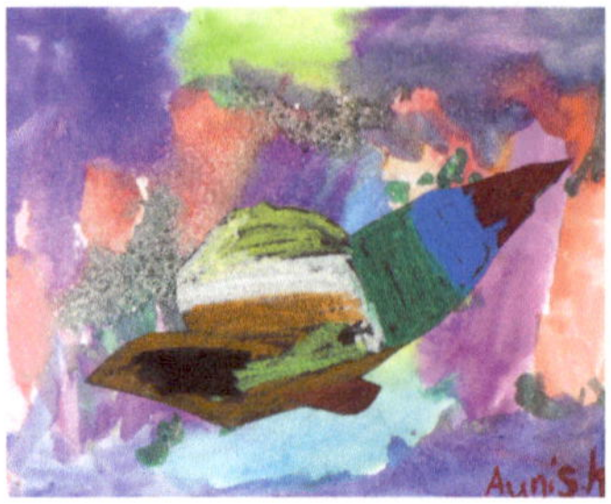
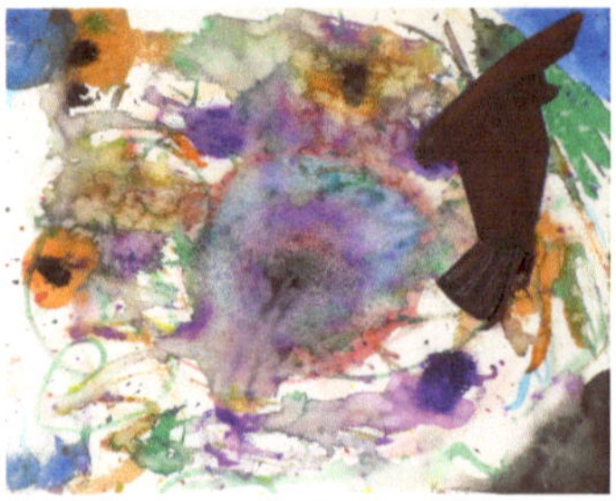
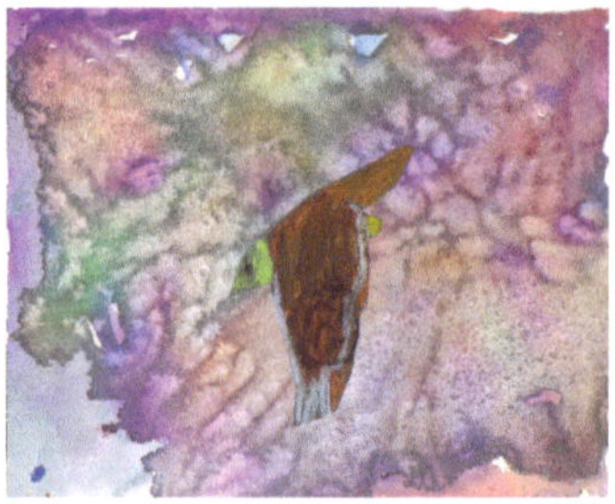
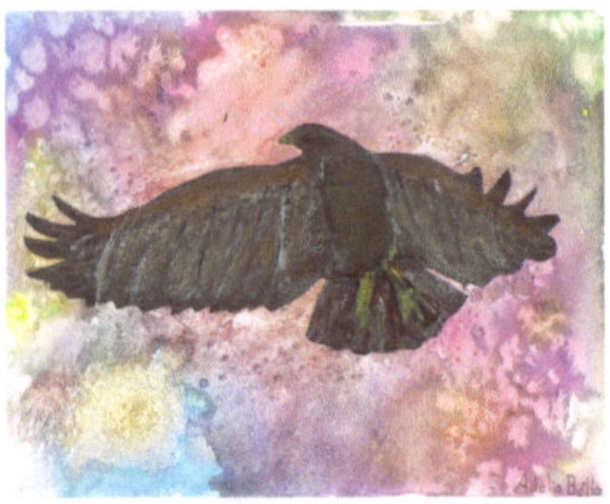
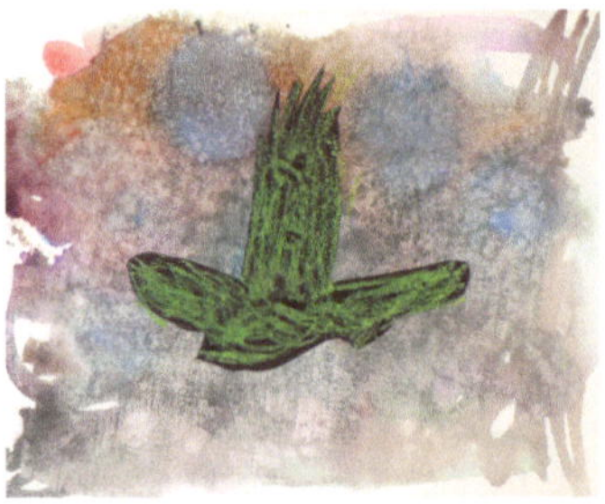
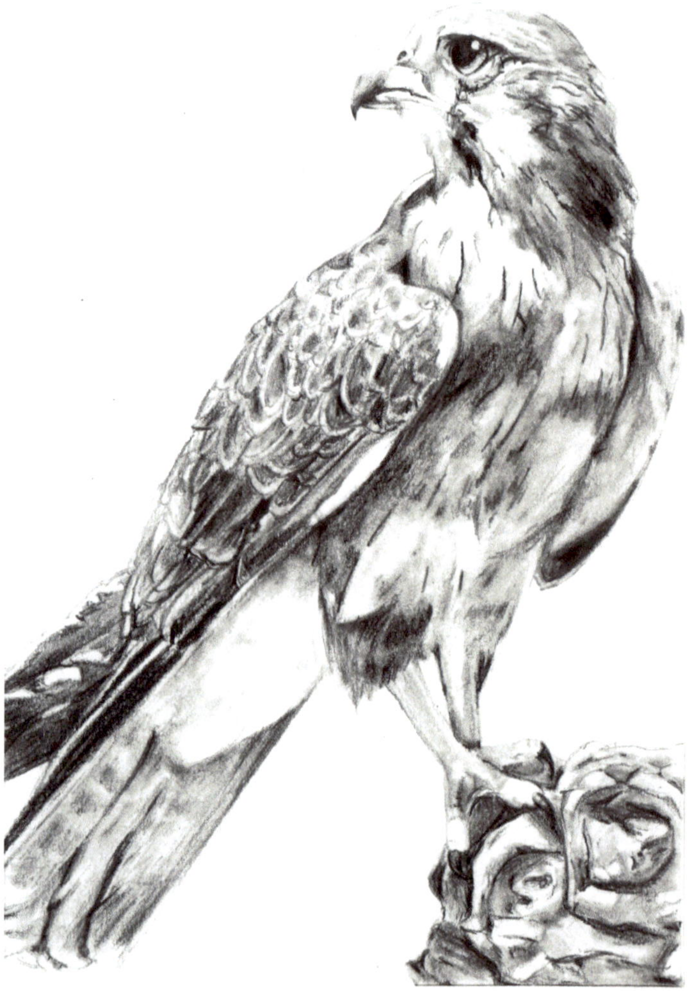
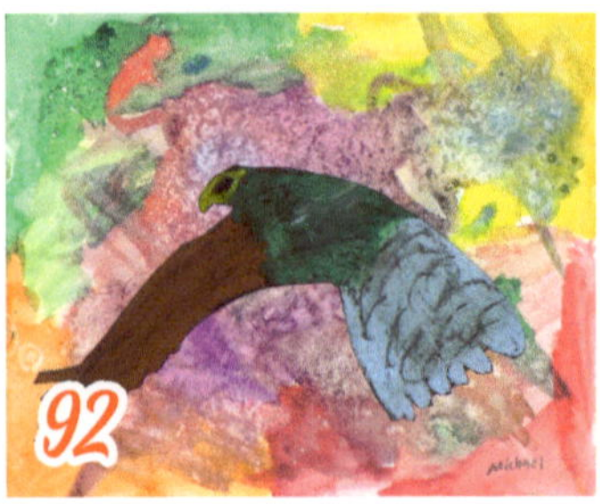
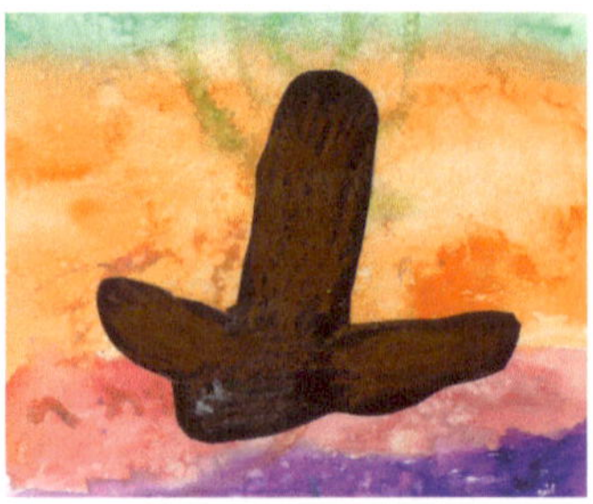
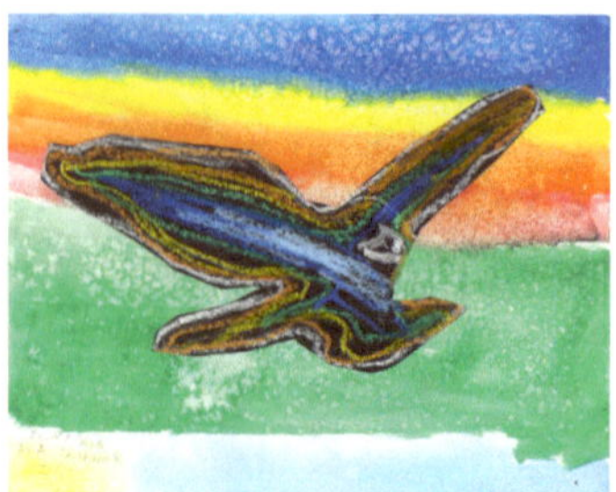
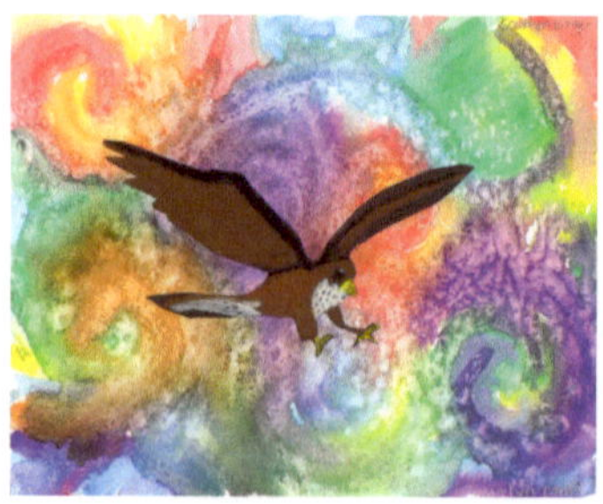

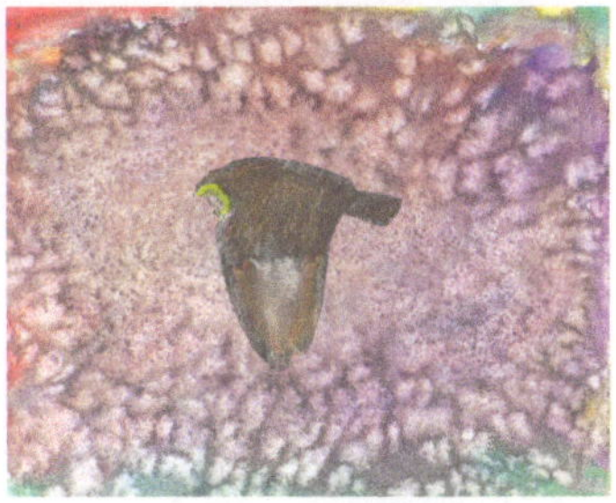
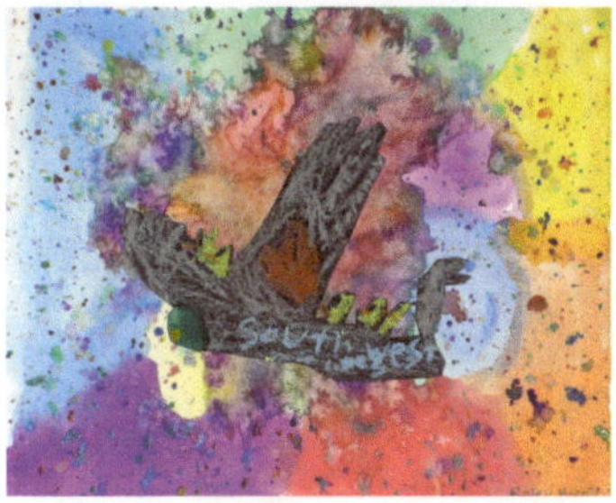
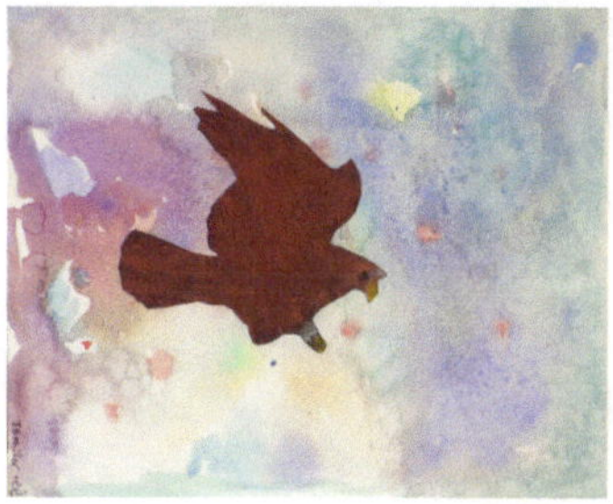
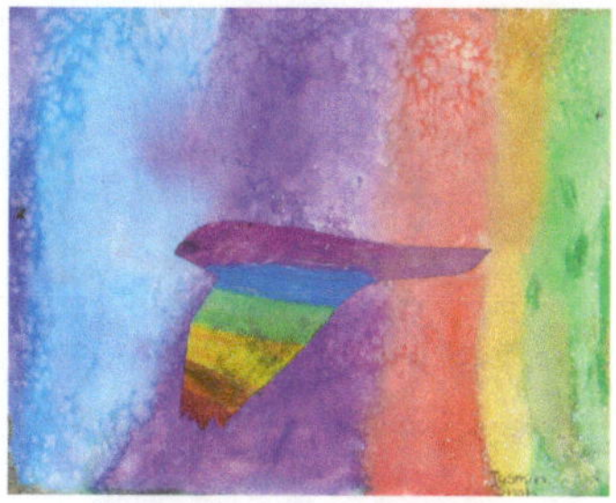
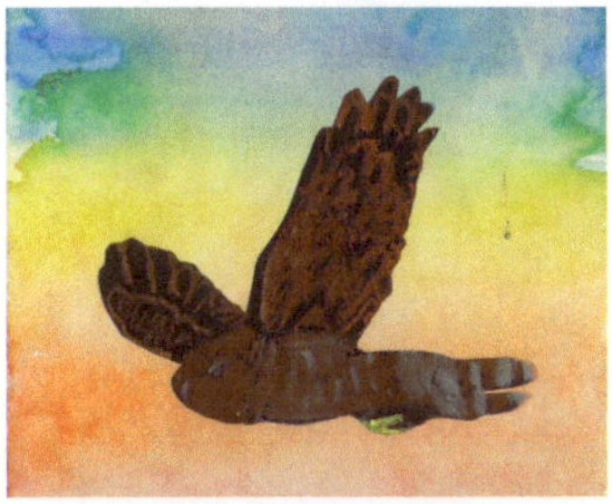
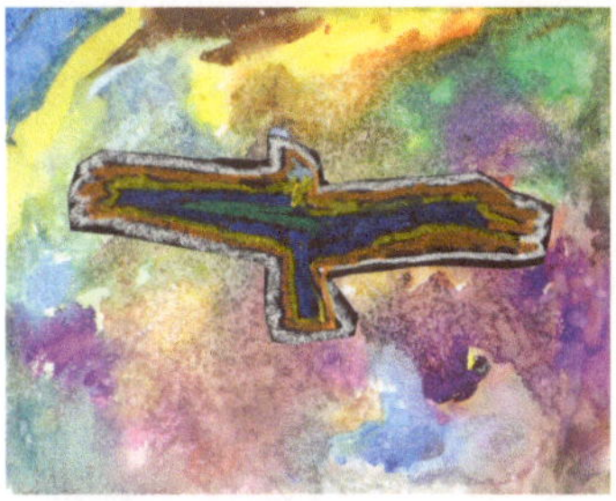
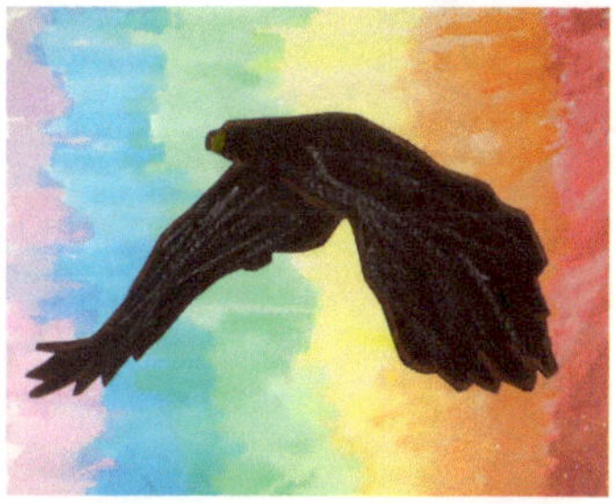
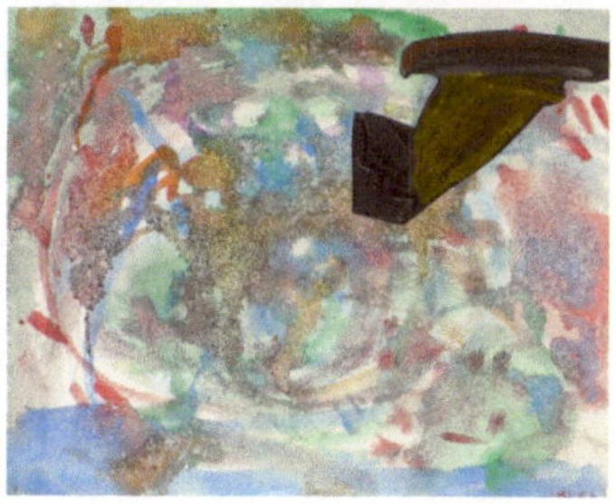
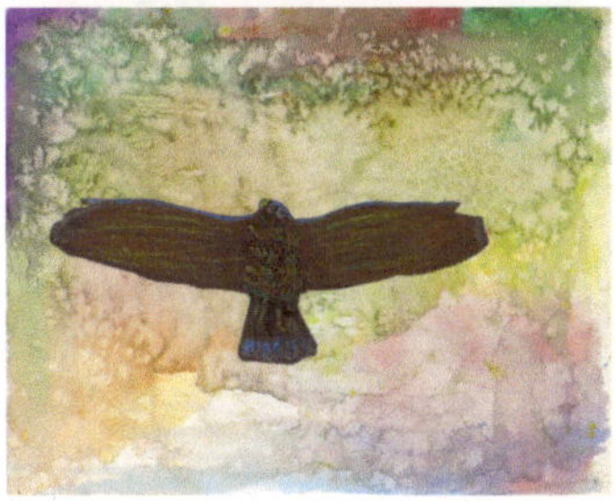
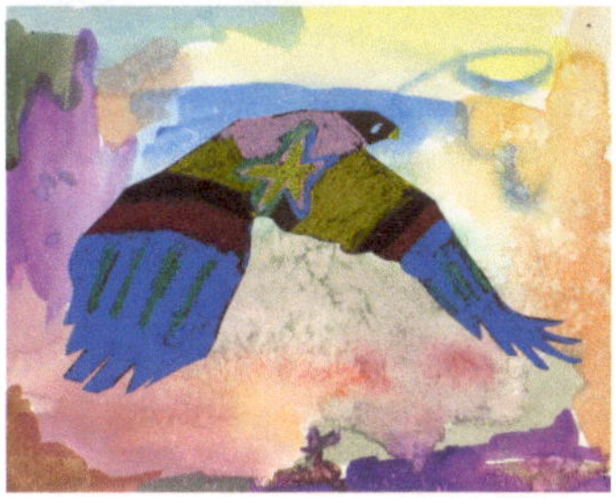
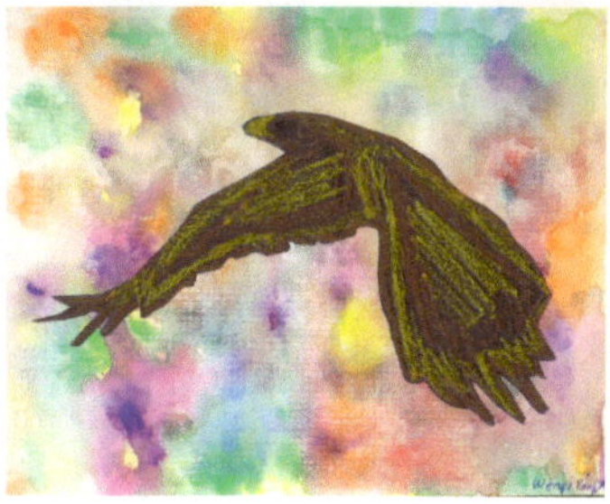
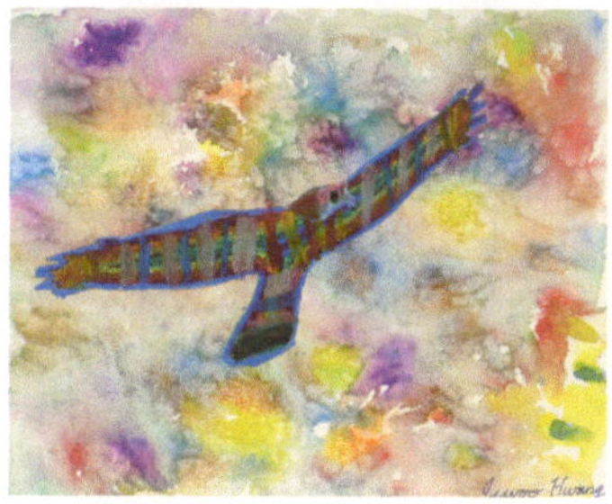
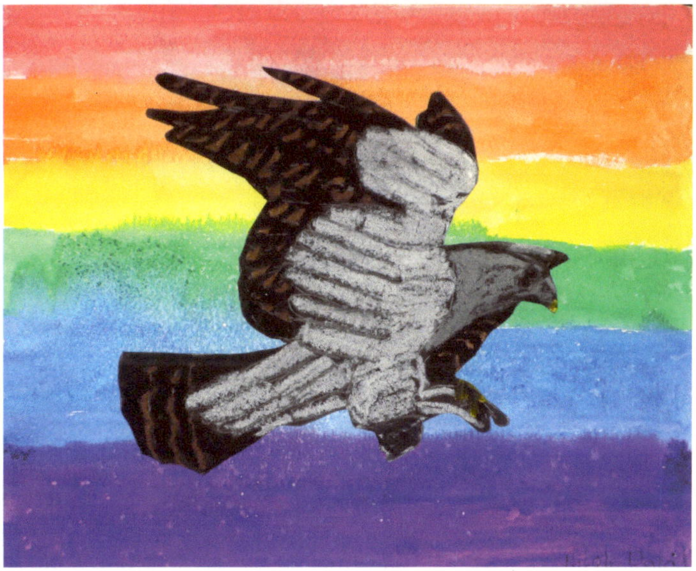
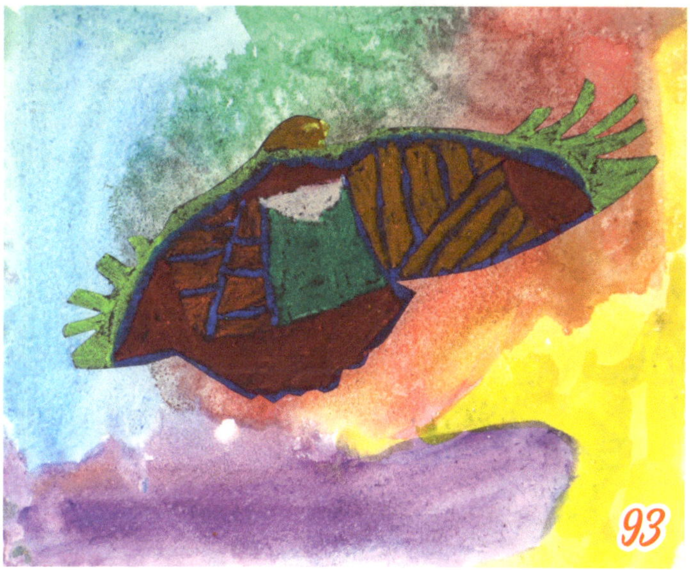

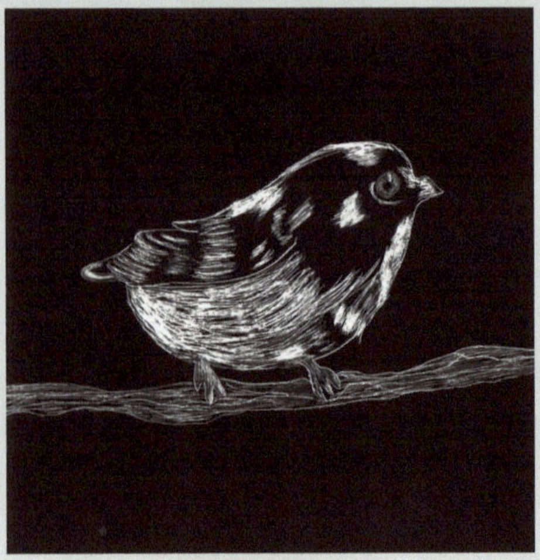

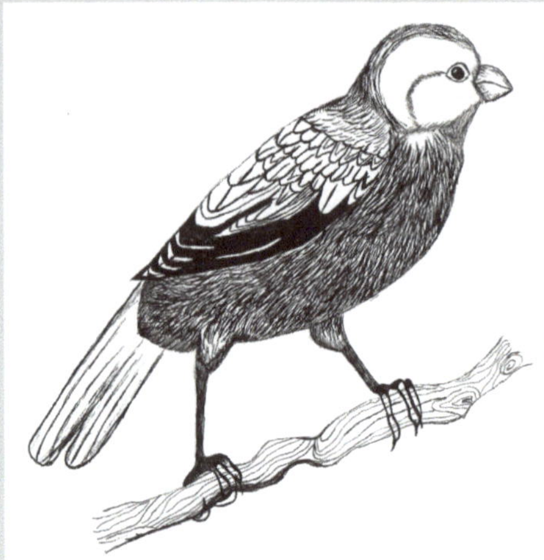

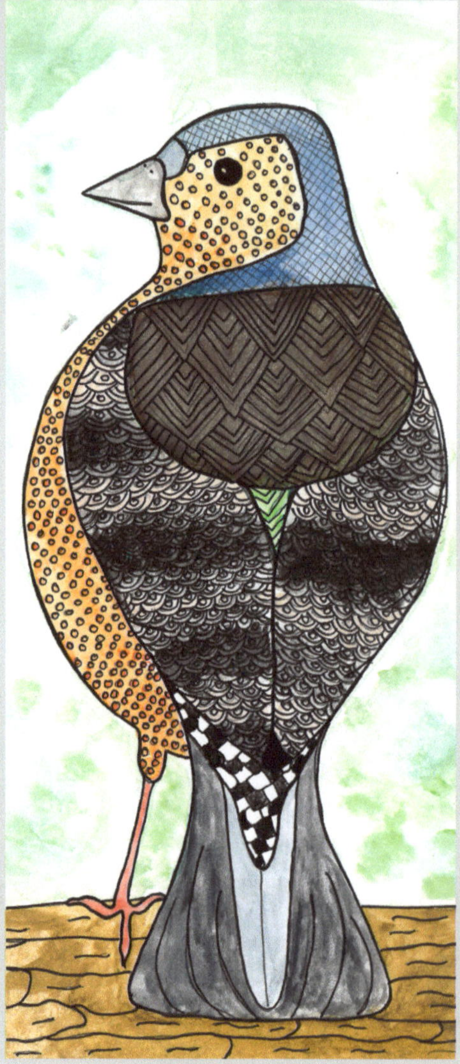

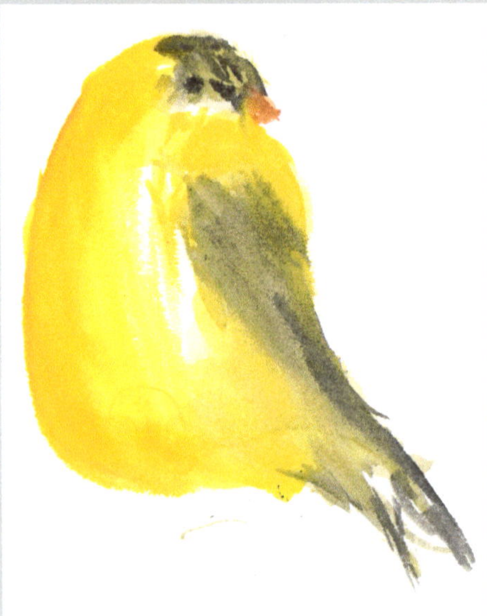

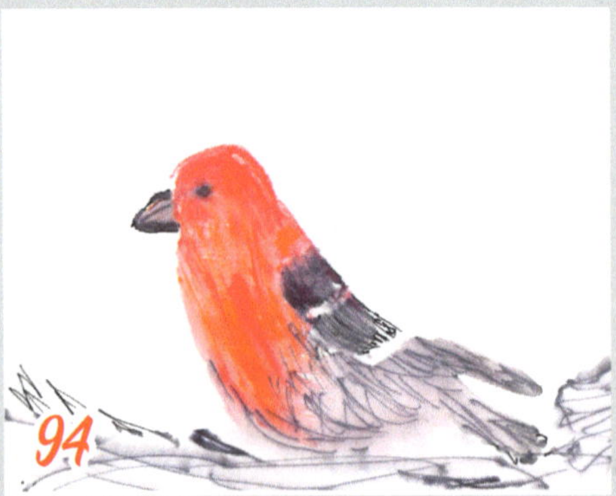

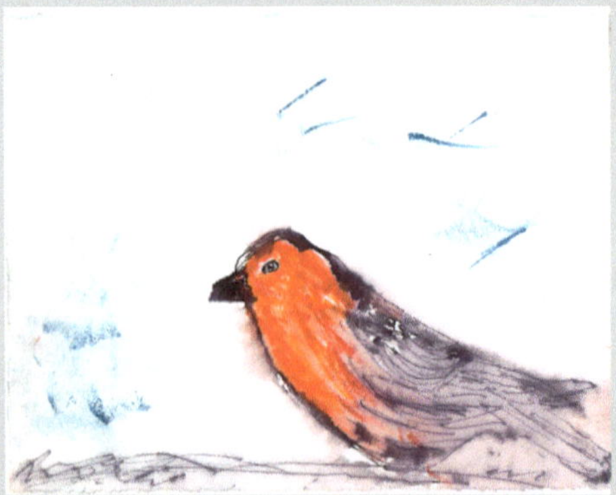

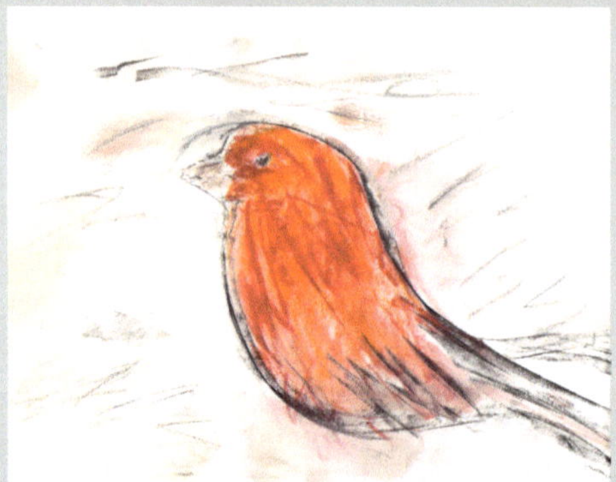

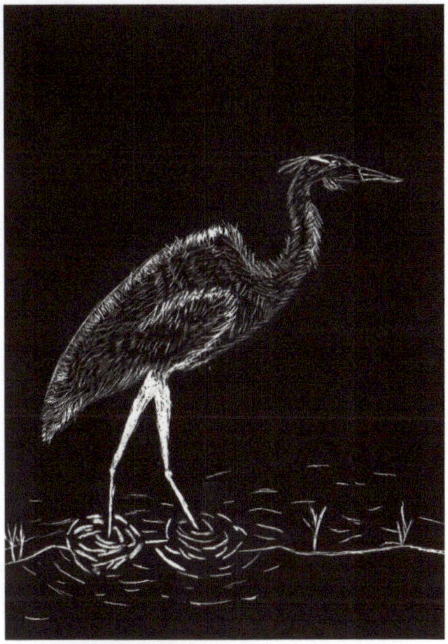

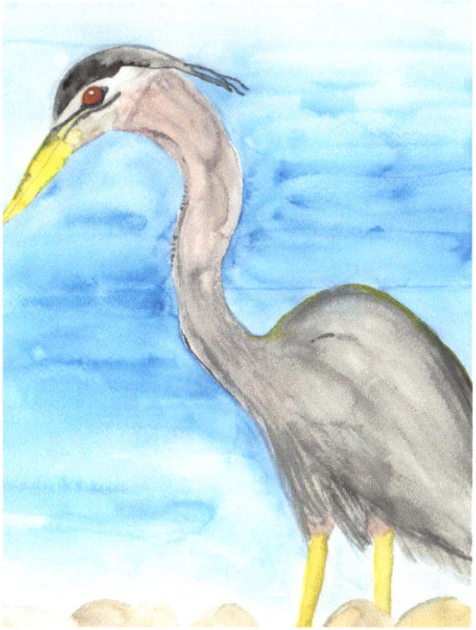

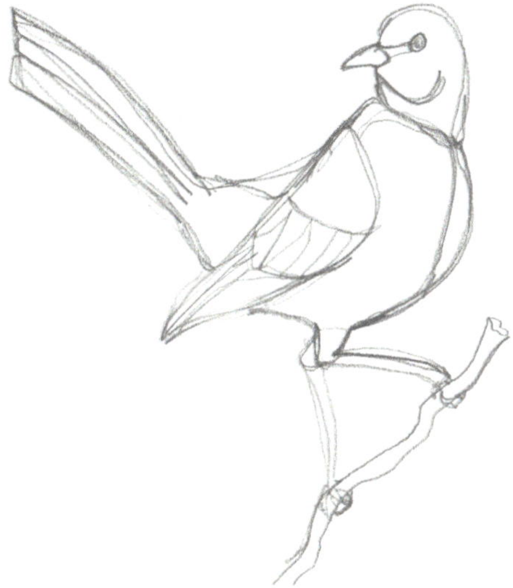

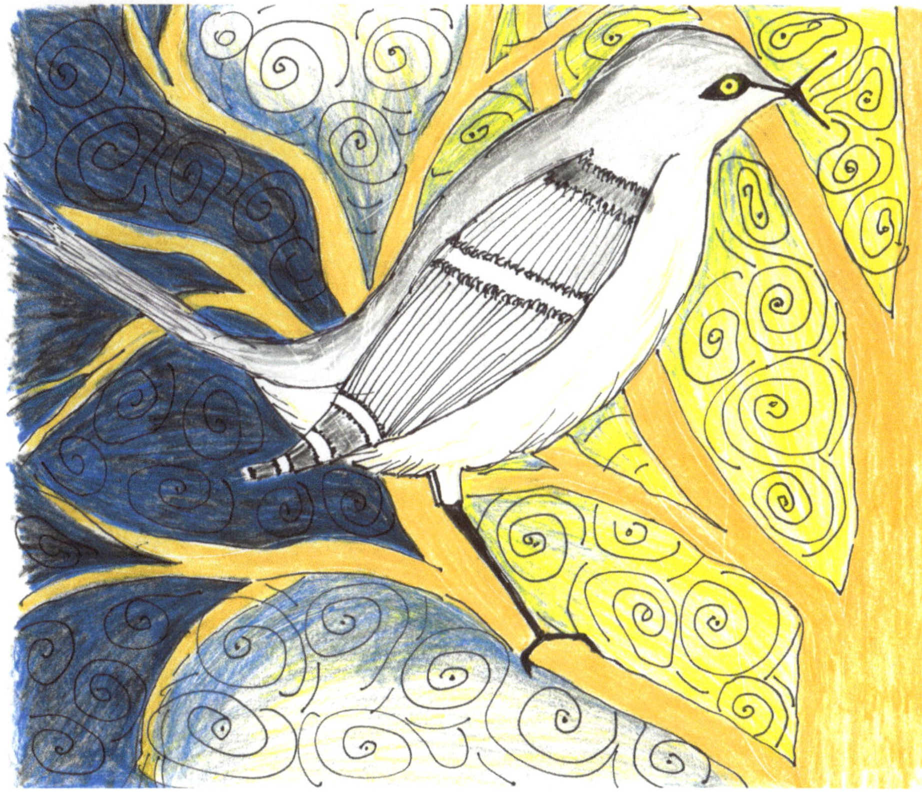

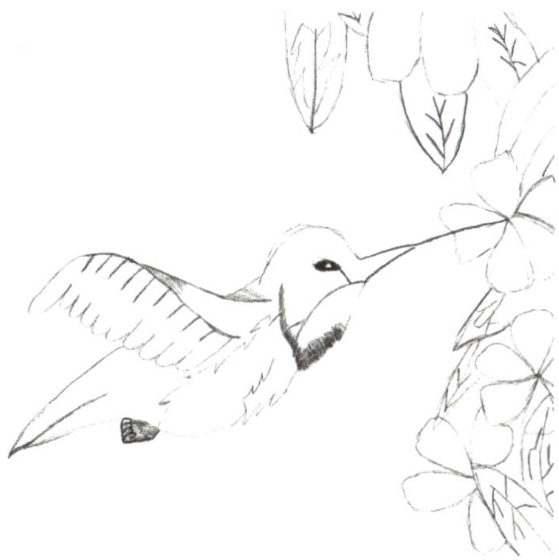

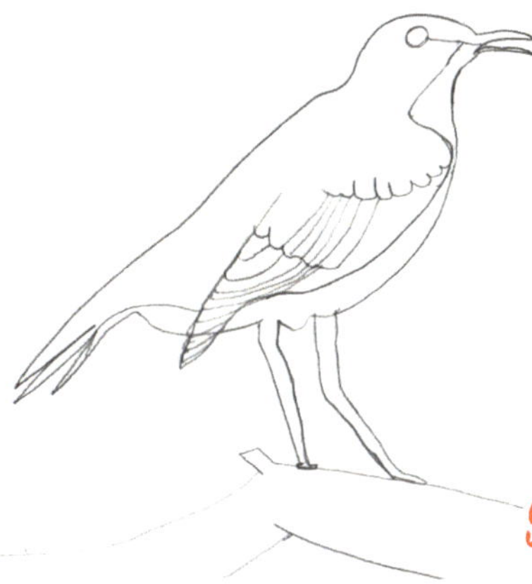

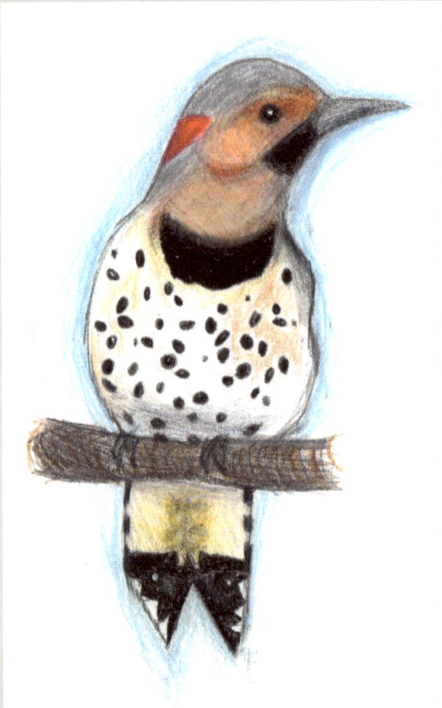
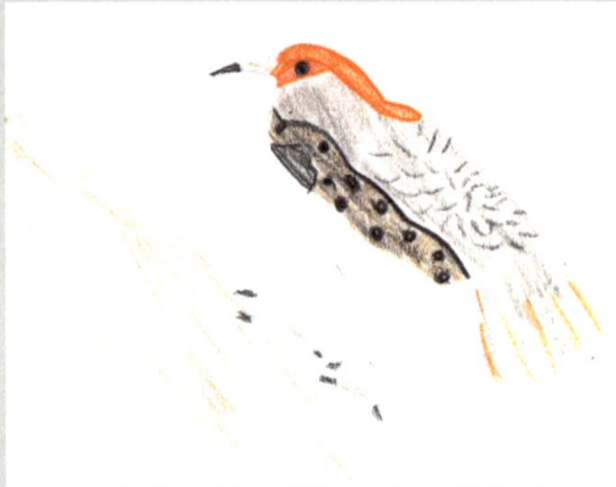
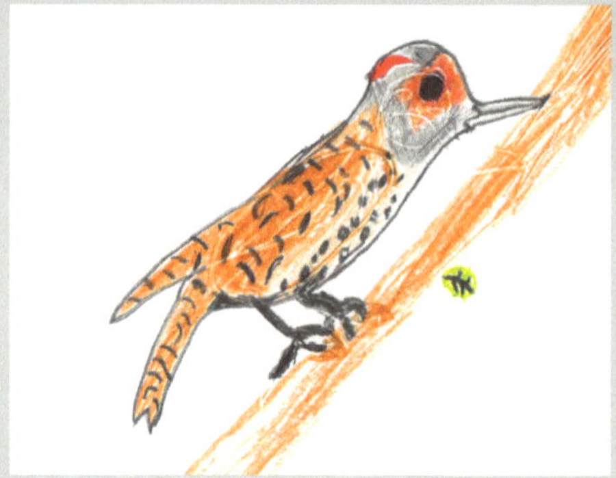
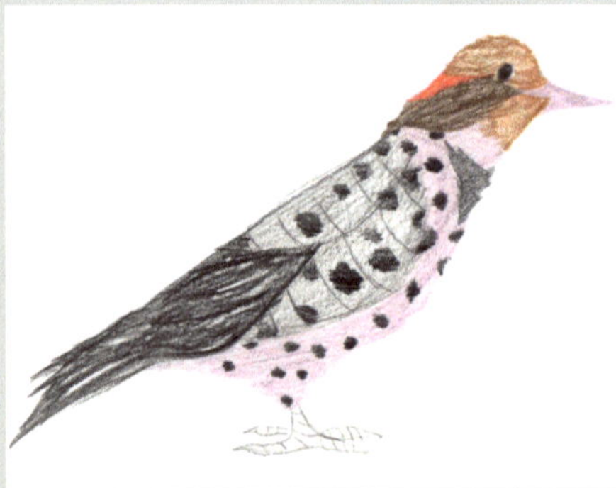
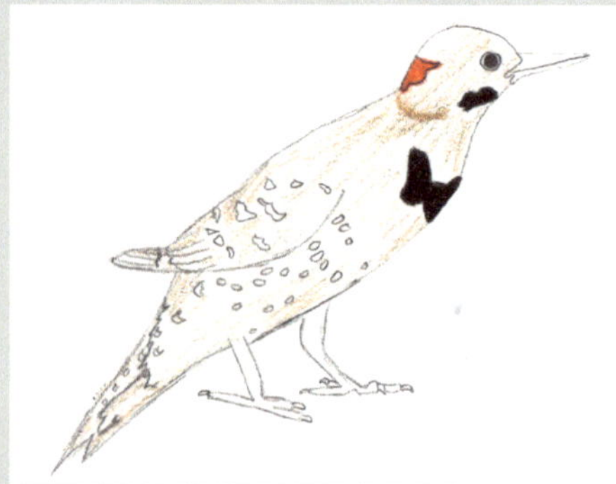
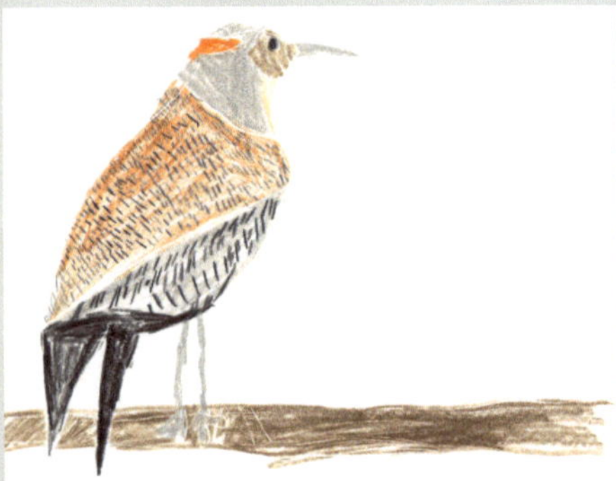
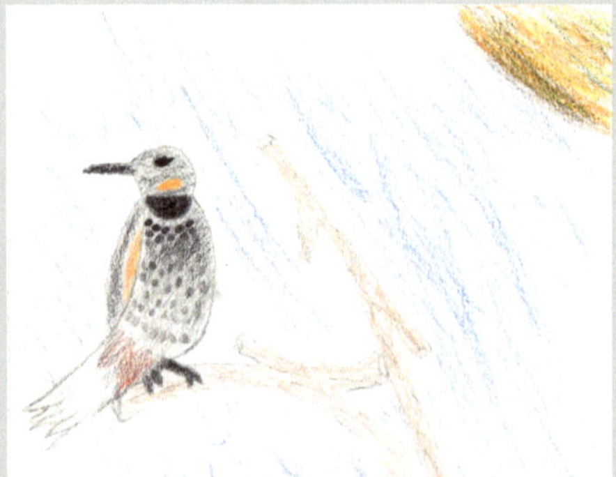
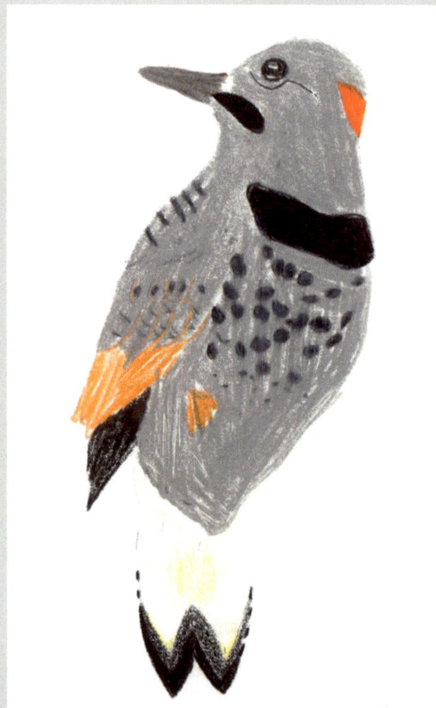
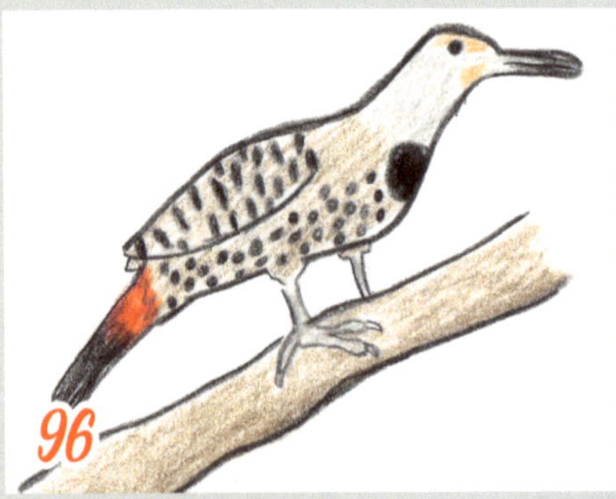
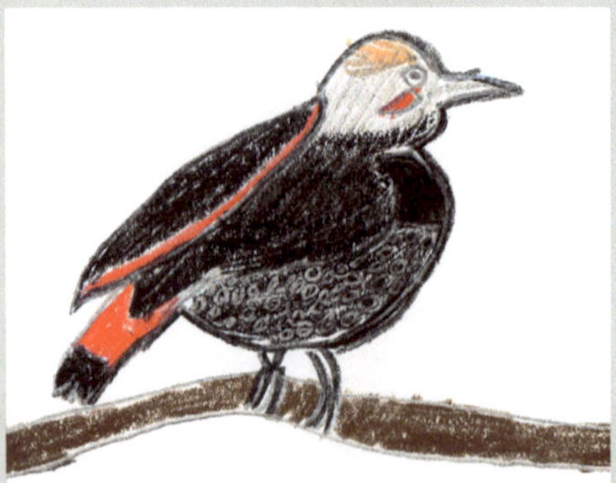

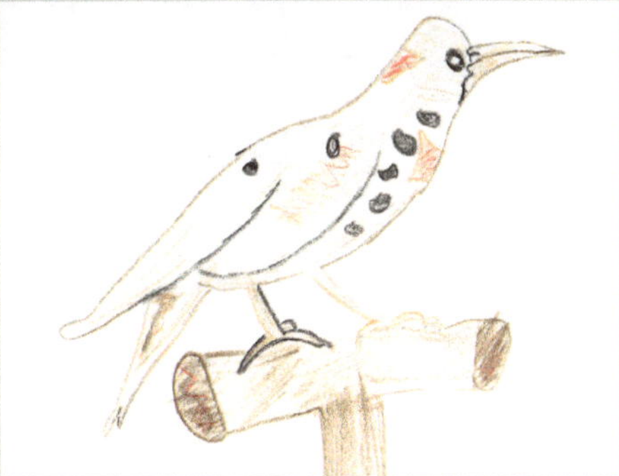
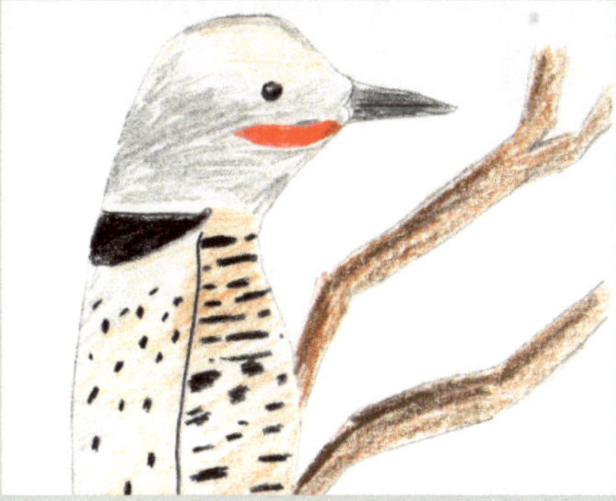
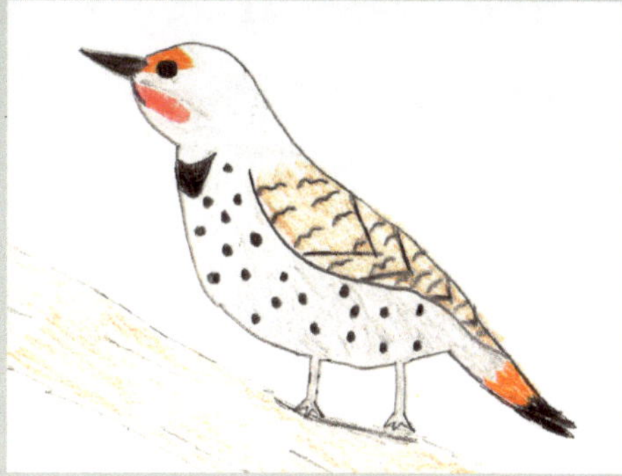
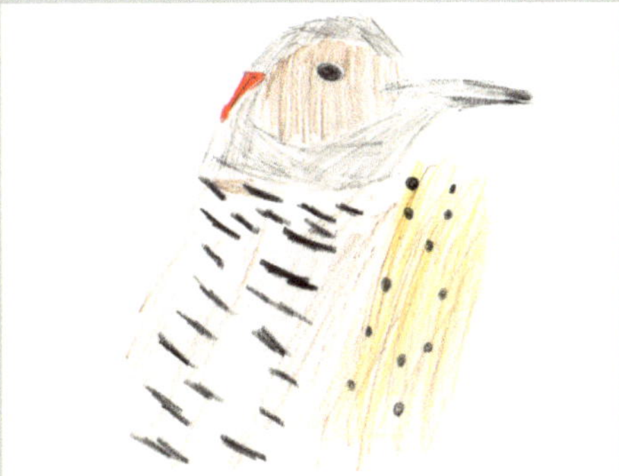
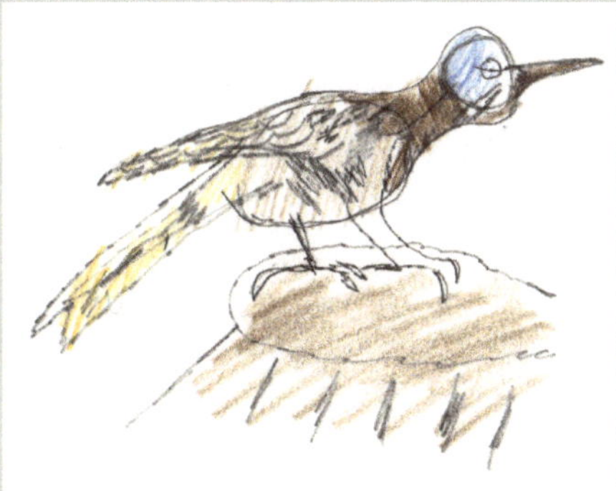
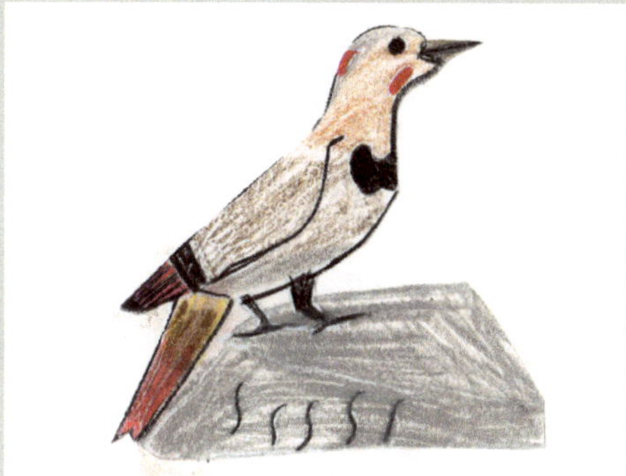
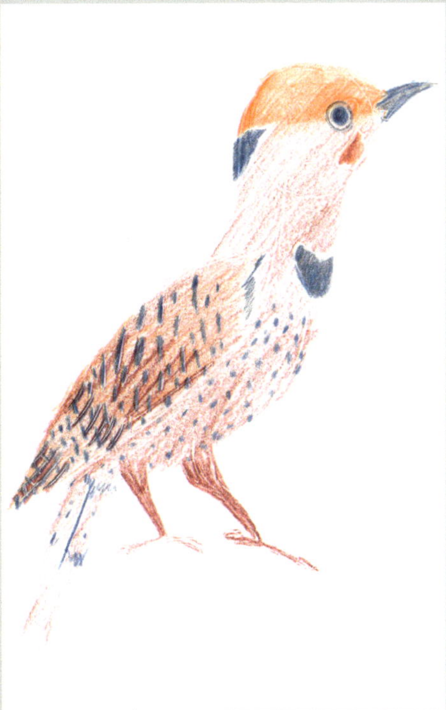
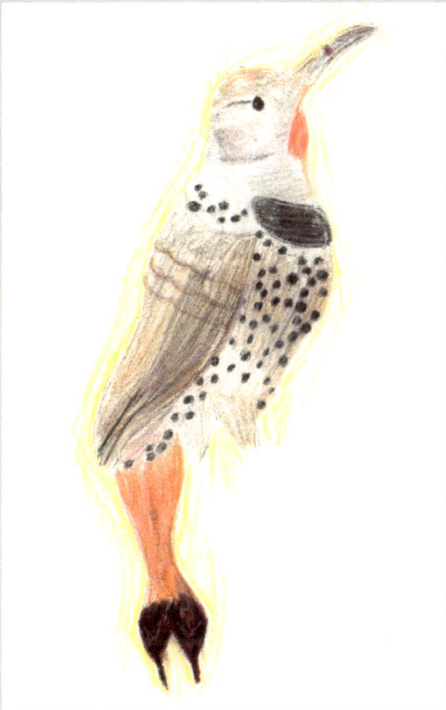
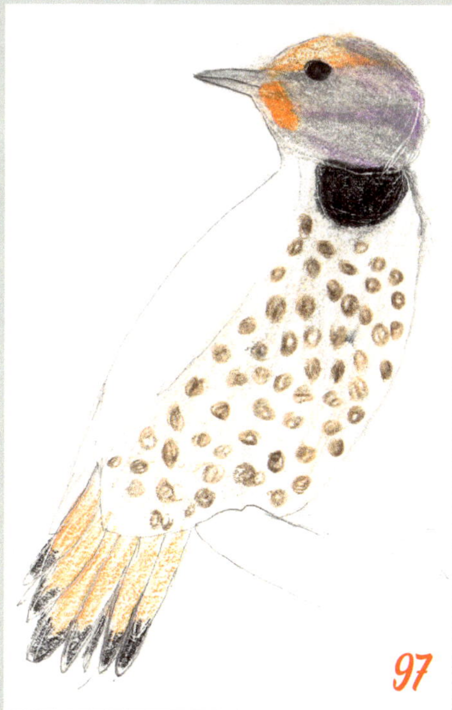

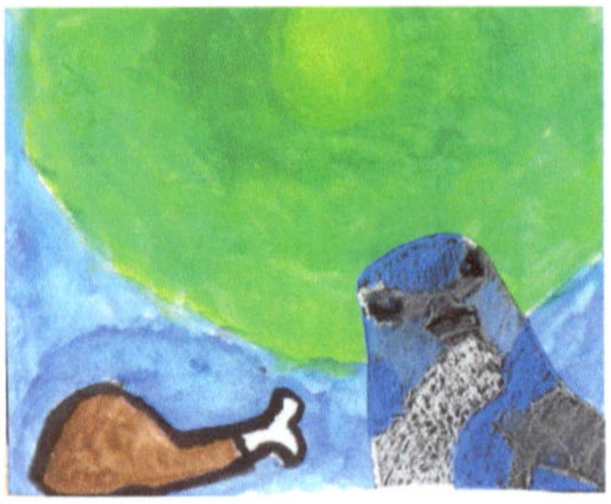
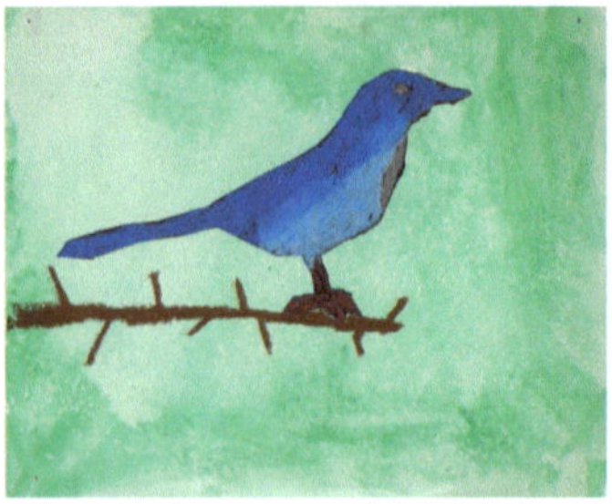
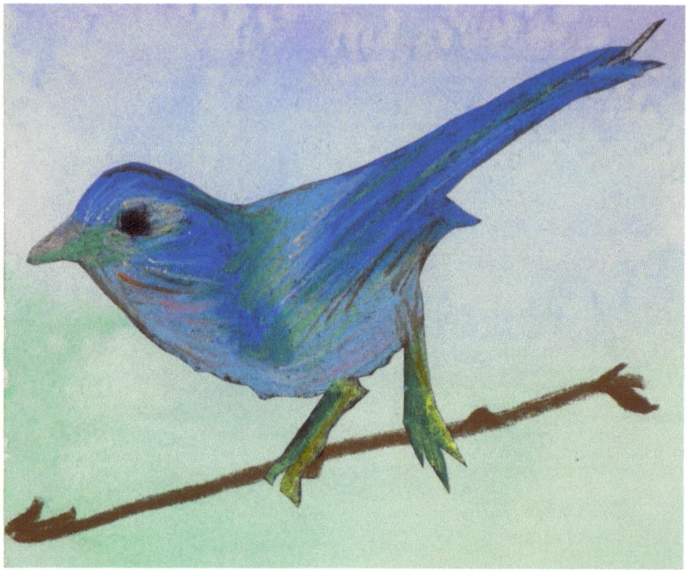
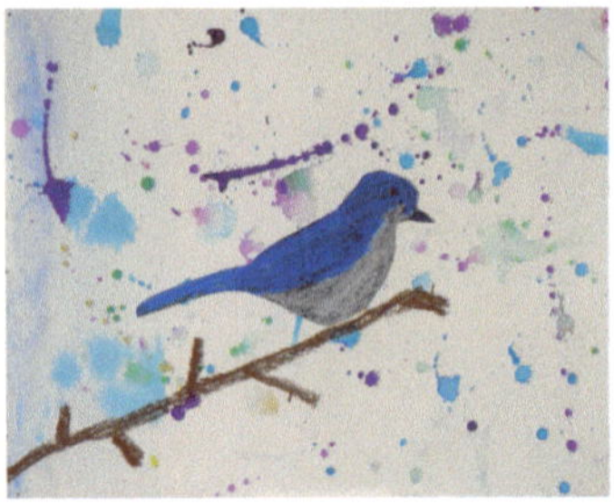
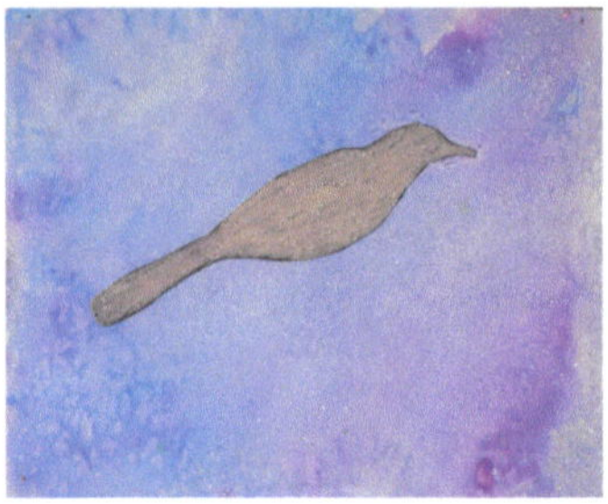
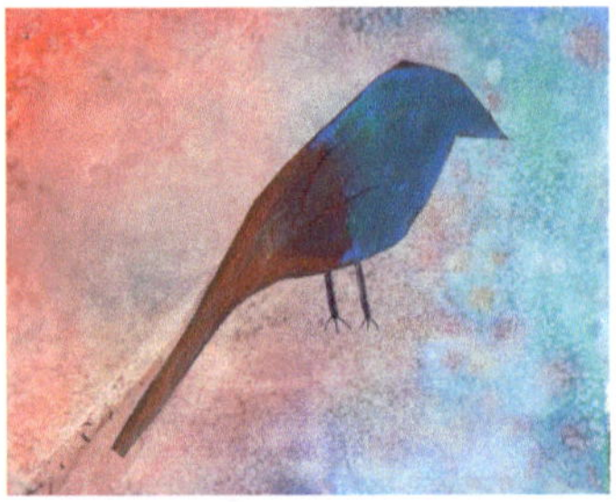
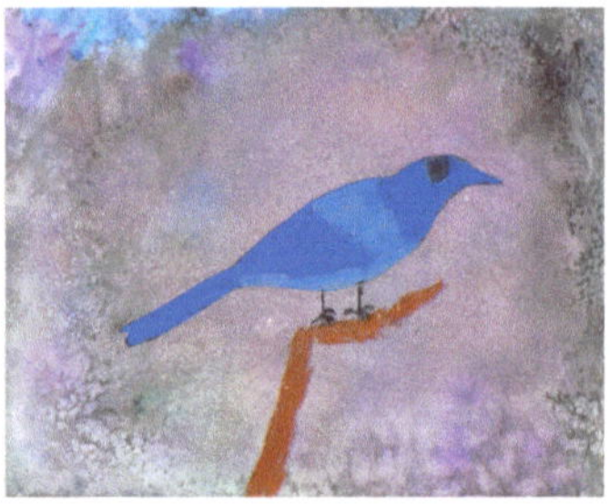
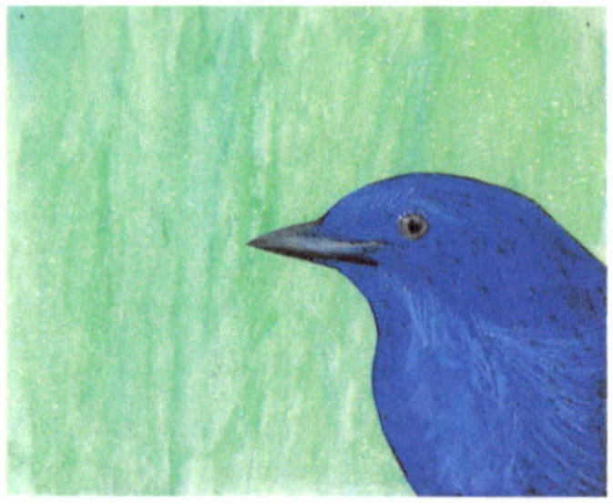
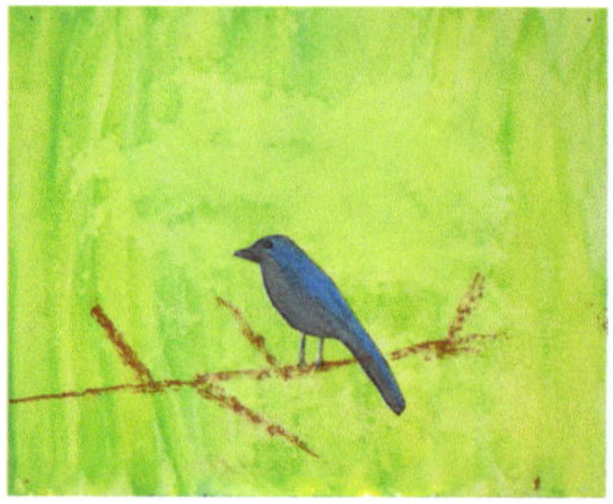
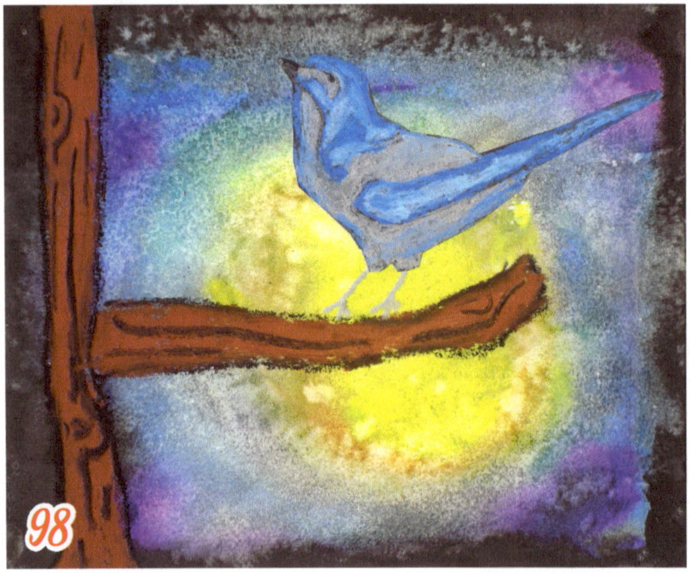
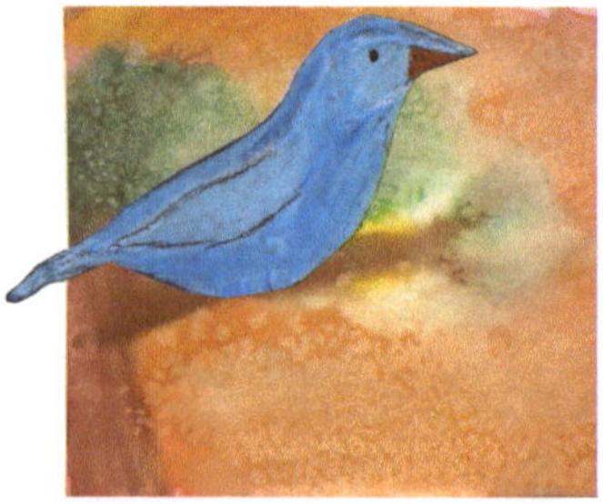
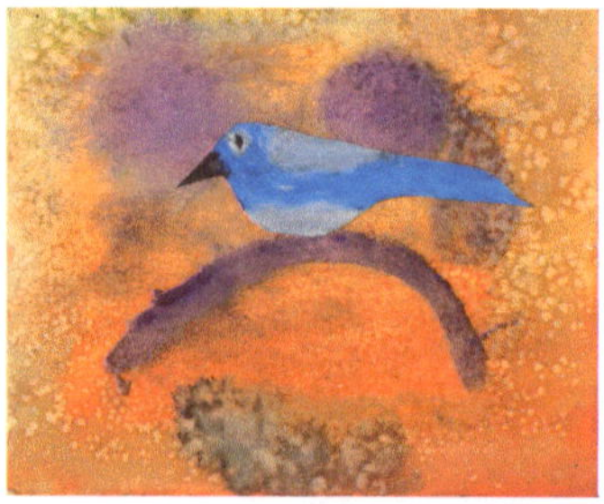
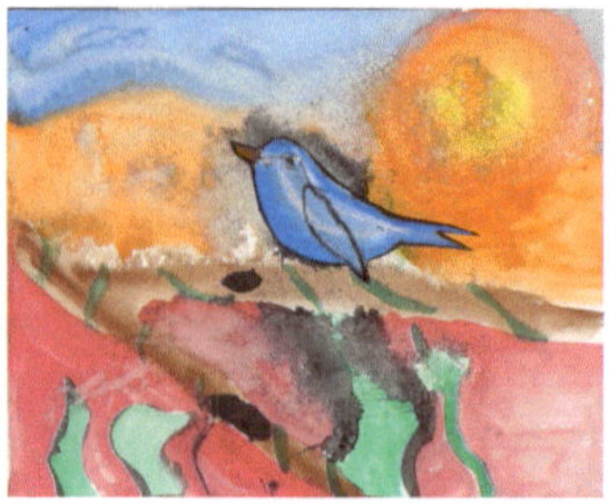
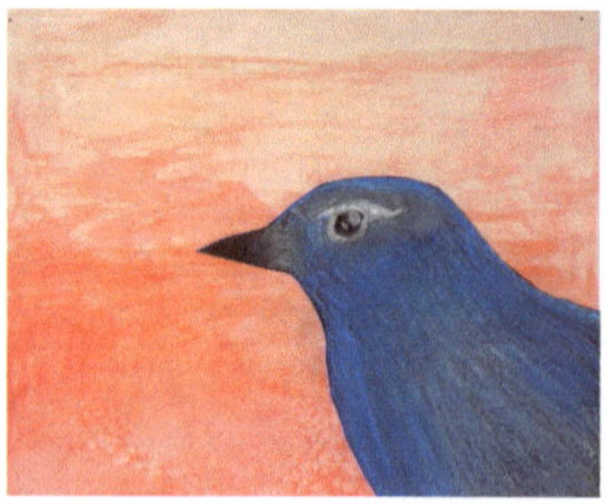

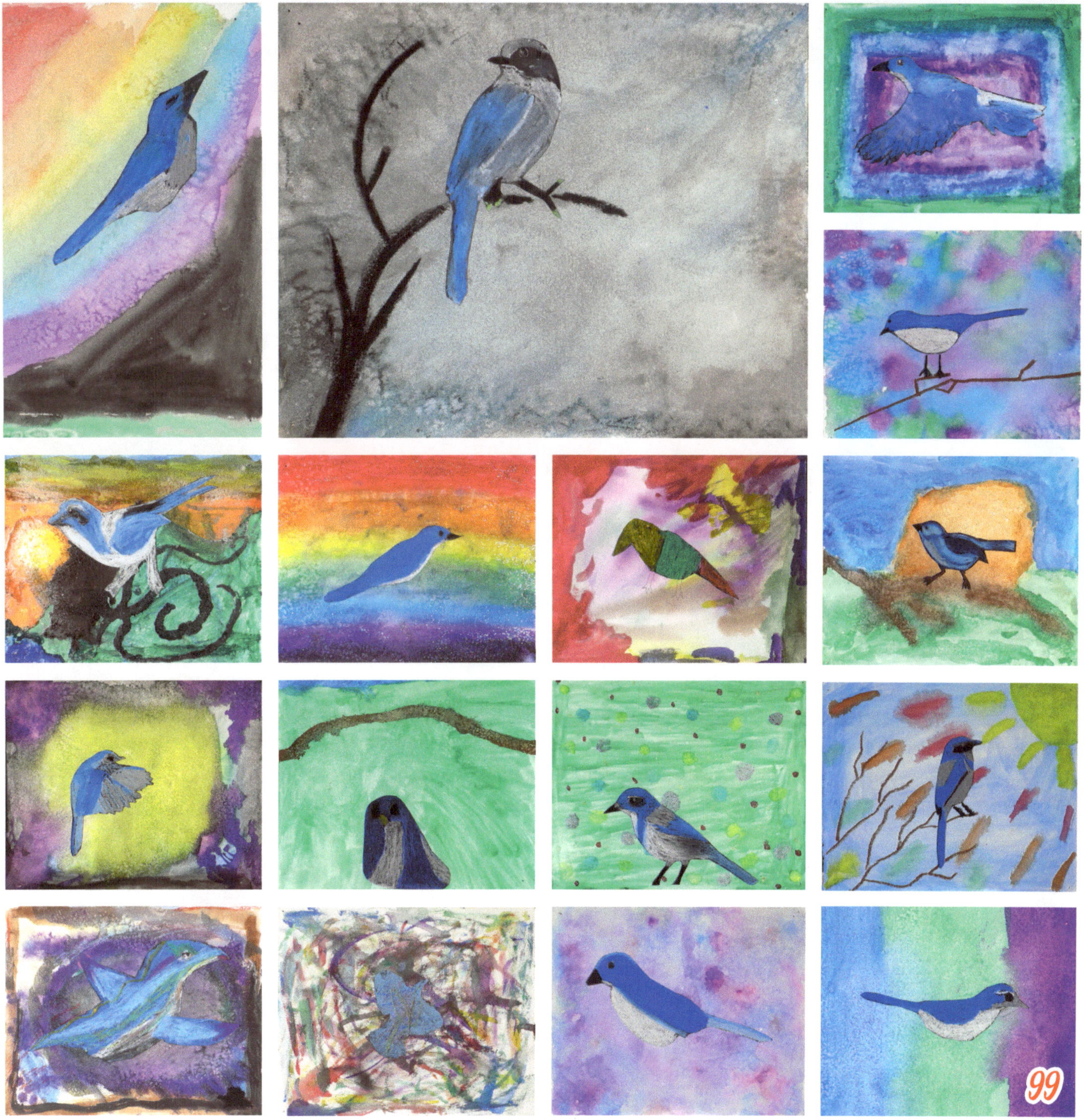

ANDSCAPE • CLOUDS • DELTA BREEZE • FLATNESS • FOG • NORTH WI

UTAH CREEK • SUNSET • PLANTS • ACORN • BERMUDA GRASS • ELDERBE

UCKEYE • BULBINE • CEANOTHUS • CENTURY PLANT • CHINESE PISTA

CHINESE TALLOW • COTTONWOOD • DAFFODIL • DATURA • DISCAR

HRISTMAS TREE • ELDERBERRY • FIG • FLOWERING QUINCE • FOX

RASS • GINKGO • HACKBERRY • HEUCHERA • IRIS • JASMINE • LEAF

EMON • LUPINE • MILKWEED • MISTLETOE • MODESTO ASH • MUST

OLIVE TREE • ORNAMENTAL PEAR • PENSTEMON • PERSIMMON • P

POMEGRANATE • POPPY • PURPLE NEEDLE GRASS • RED HOT POK

EDBUD • REDWOOD • ROSEMARY • SALVIA • SOURGRASS • SUNFLOW

CAMORE • TOMATO • TOYON • TULE • VALLEY OAK • WALNUT • FISH A

EPTILES • FROG • SALMON • TURTLE • MAMMALS • CAT • COW • COYO

OG • HORSE • OTTER • OPOSSUM • RABBIT • RACCOON • RAT • SQUIR

NSECTS • BEE • BLACK WIDOW SPIDER • BUTTERFLY • COCKROA

RANE FLY • DEERFLY • DRAGONFLY • FLY • MOSQUITO • MOTH • PRAY

ANTIS • FANTASY • BIRDS • BLACKBIRD • CANADA GEESE • CHIC

CROW • DOVE • DUCK • EAGLE • EGRET • FINCH • FLICKERS • HAV

ERON • HUMMINGBIRD • MAGPIE • MOCKINGBIRD • OWL • SC

Y • STARLING • SWALLOW • TANAGER • TURKEY • TURKEY VULT

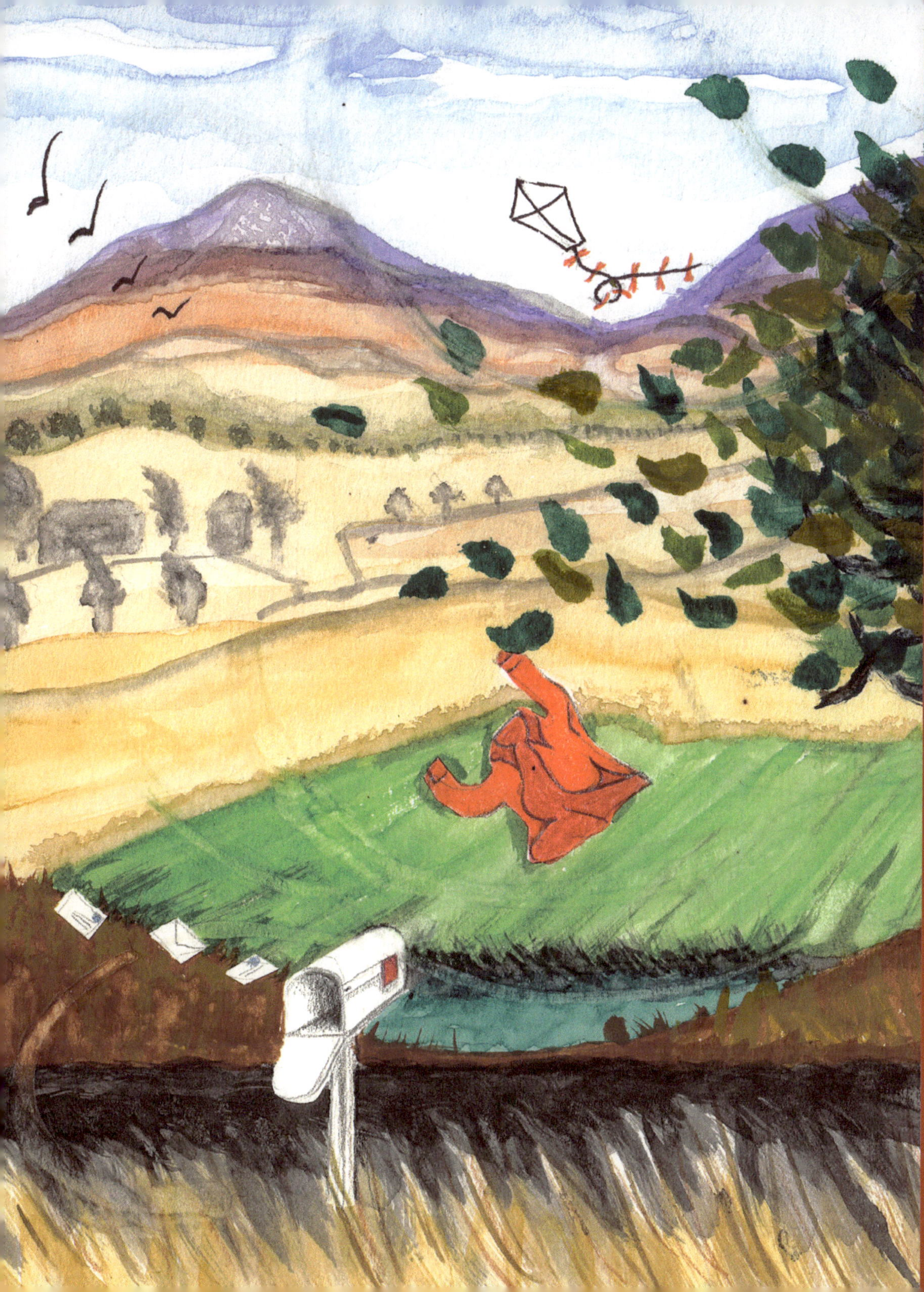

LANDSCAPE

The following artists created the artwork in this section. The list is alphabetical by category and does not necessarily represent the order nor page number the artwork appears; please refer to the page numbers next to the category to locate where the art appears. Some teachers/schools did a class project (CP) for one category; all the artists involved in that project are listed in italics under the teacher and school name within the category. For other class projects the artist chose their category; their names appear under that category with a footnote indicating their teacher/school. If there is a number after an artist's name they provided more than one piece of art for that category.

CLOUDS
(pages 103-105)
DeLeon, Christina - CP[1]
Voss, Angela - CP[2]
CP, 2nd, Elia Chavez,
Art Teacher Jami
Rubin-Mark, Cesar
Chavez Elementary
Brenner, Eli
Deering, Ruby
Dotson, A'riyah
Ghassemi, Mimi
Grey, Agatha
Hales-Young, Lilian
Heiser, Whitney
Hill, Velia
Incendy, Ty
Kerr, Emma
Levin, Epenai
Lin, Andrew
Mark, Gabriel
McCourt, Leighton
Rogers, Juniper
Ross, Ramon
Santoscoy, Iliana
Simmons, Alittie
Stallard, Sebastian
Thompson, Finley
Wallace, Kiera
Wollrich, Leah
Wong, Dante
Young, Luke

DELTA BREEZE
(pages 110)
Sherwood, Barbara
Williams, Judith

FLATNESS
(page 106)
Do, Kimberly
Sublett, Carol
Williams, Judith

FOG
(page 107)
Ryland, Rebecca

NORTH WIND
(pages 101 and 107)
Moores, Judith
Taylor Tuft, Candy
Wagner, Jerri

PUTAH CREEK
(page 107)
Ryland, Rebecca

SUNSET
(pages 108-109)
Barbosa, Ensley - CP[2]
Brady, Susan Whitley
Burgess, Jaclynn - CP[1]
Palchak, Michael - CP[2]
Salazar, Suzanne - CP[1]
Sullivan, Jeannette - CP[1]

[1]Student's Choice Class Project, Teacher: Melanie Bown, St. Francis High School • [2]Student's Choice Class Project, Teacher: Carrie Ceniseroz, St. James School

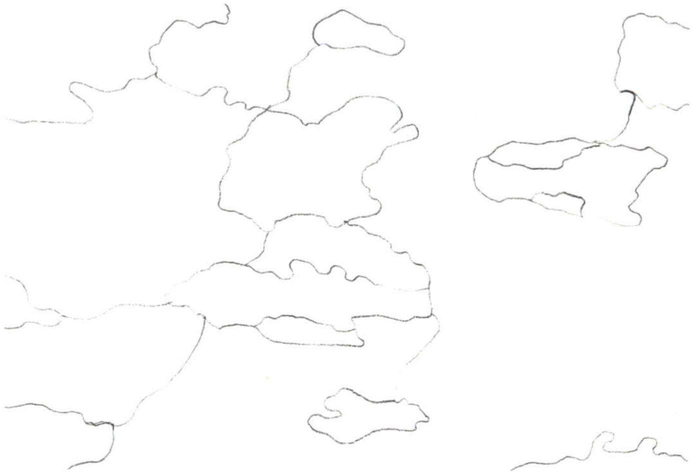

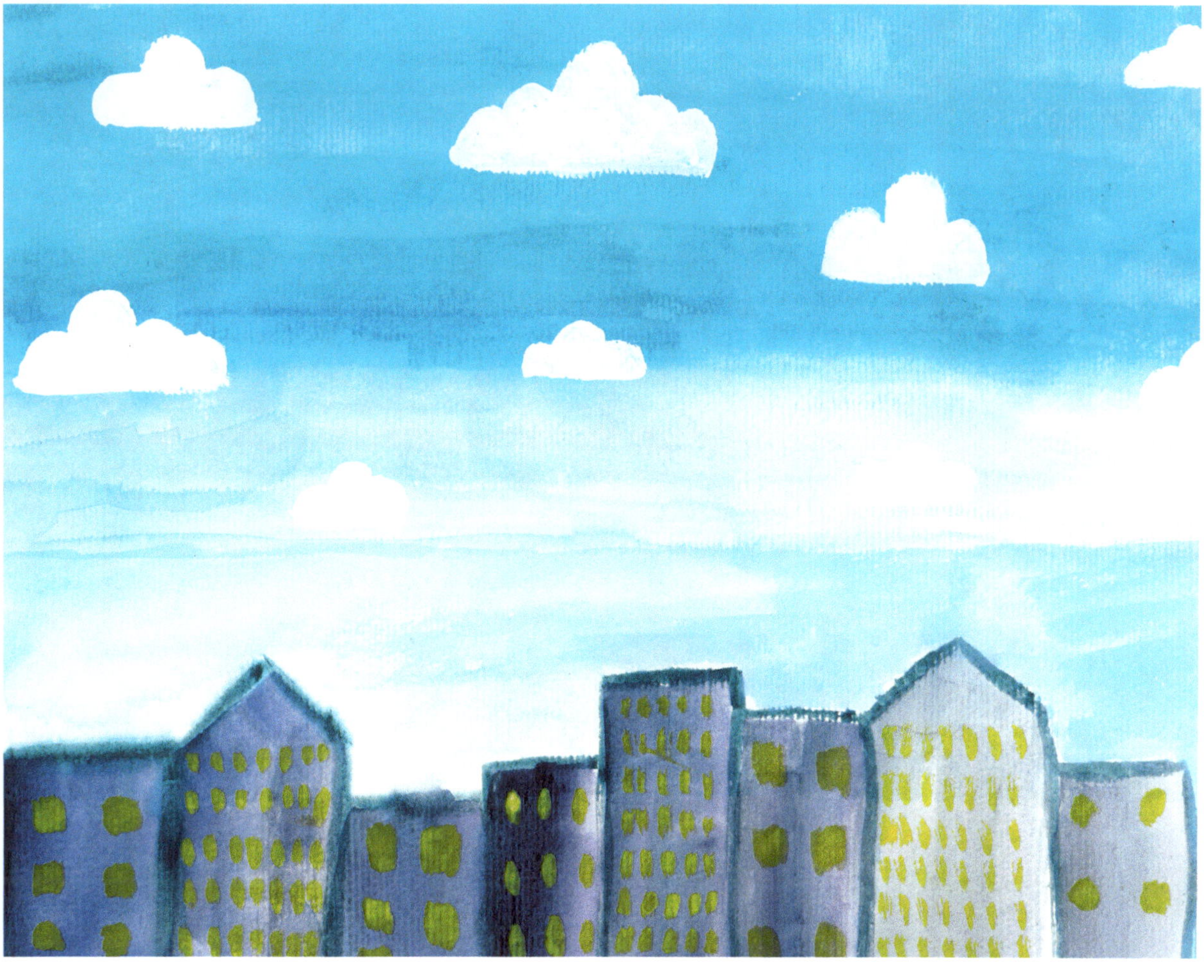

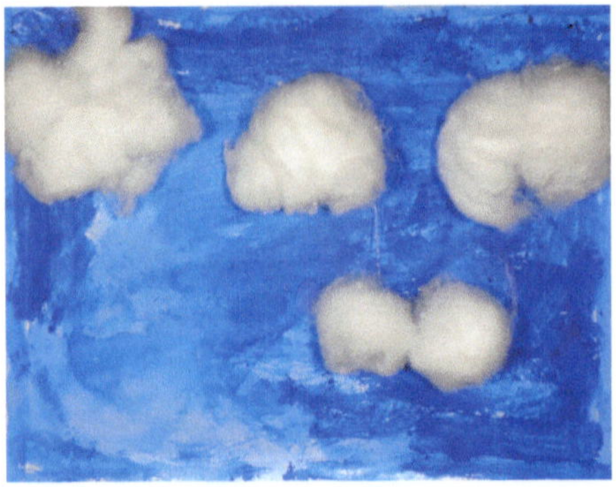
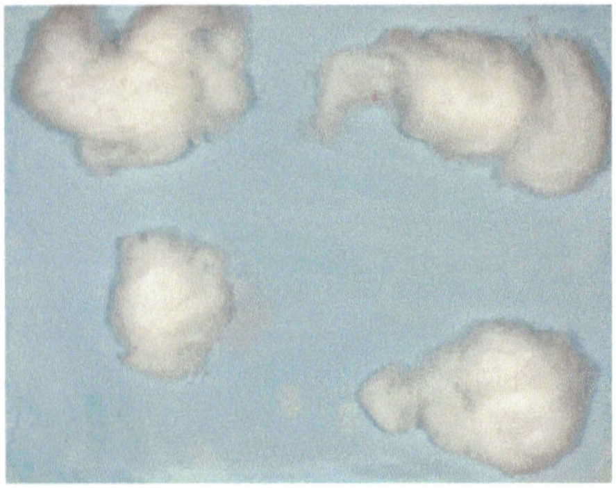
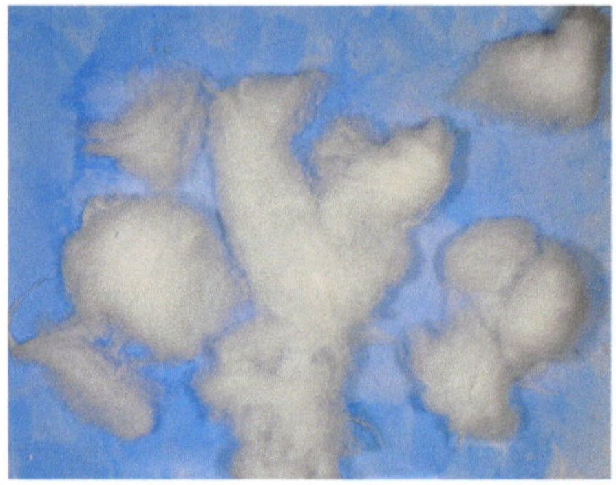
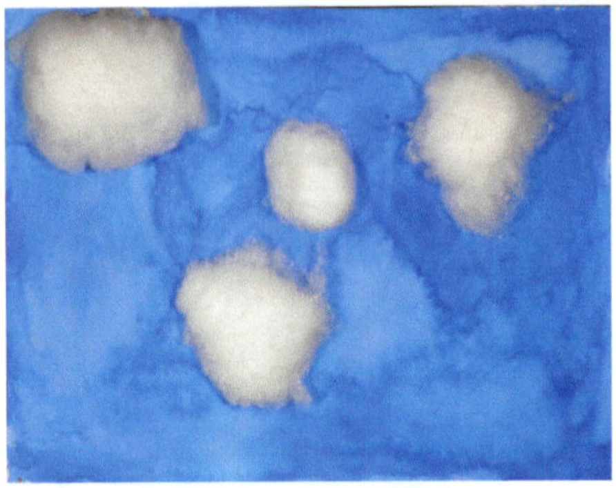
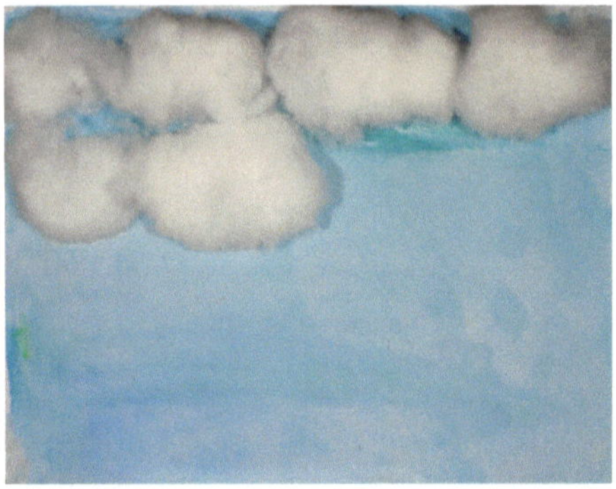
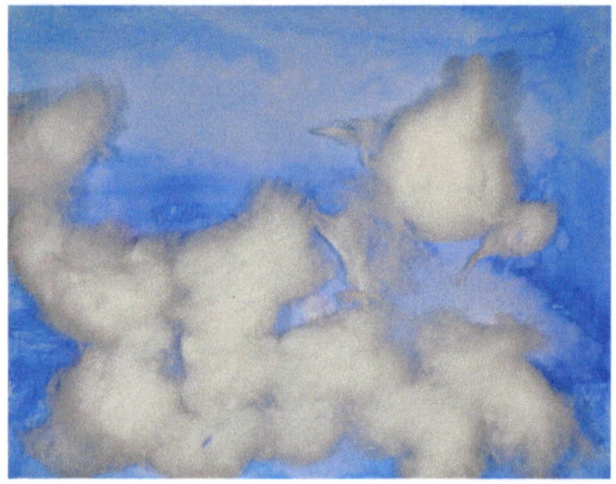
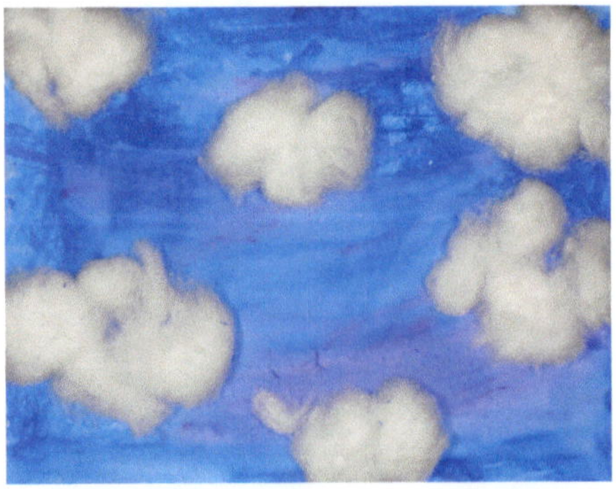
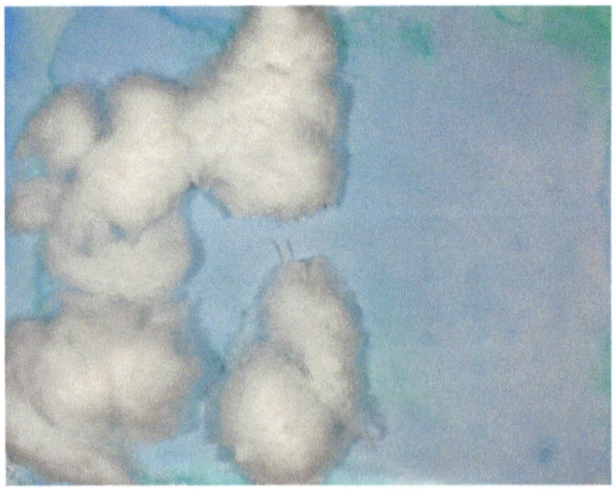
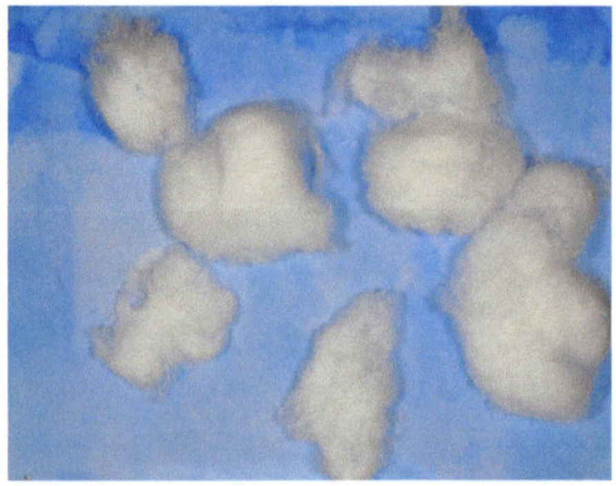
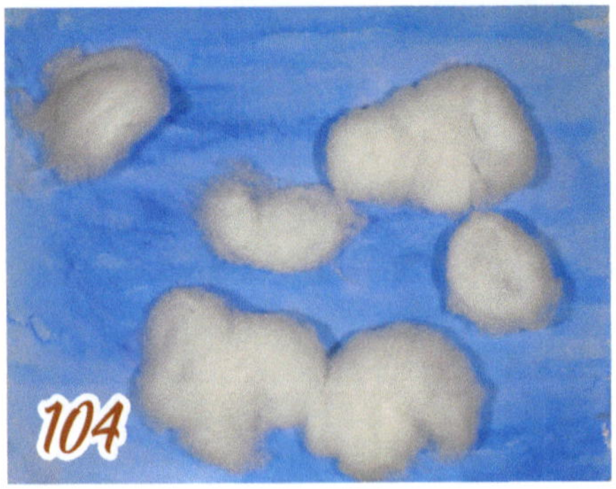
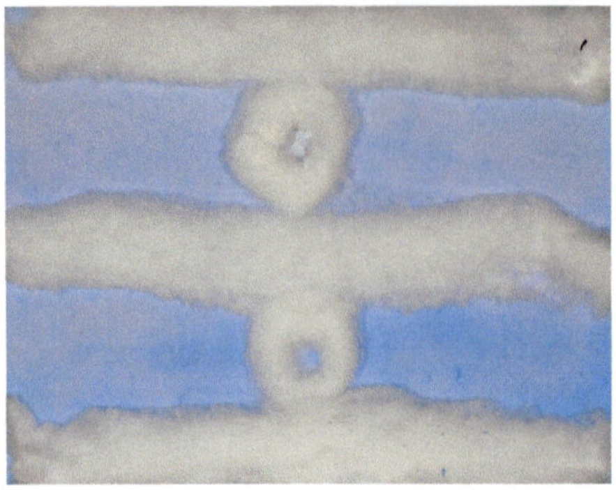
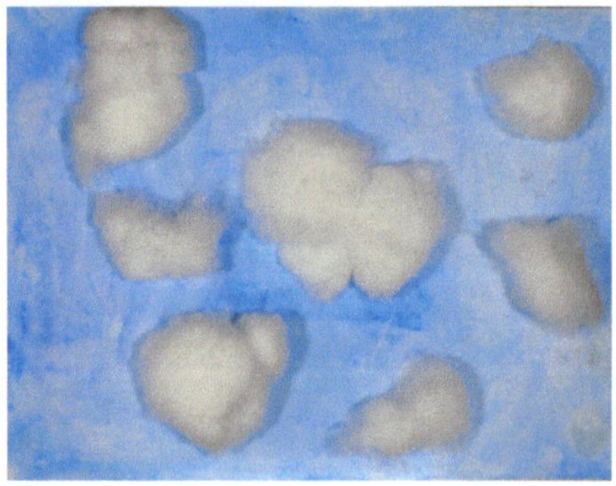

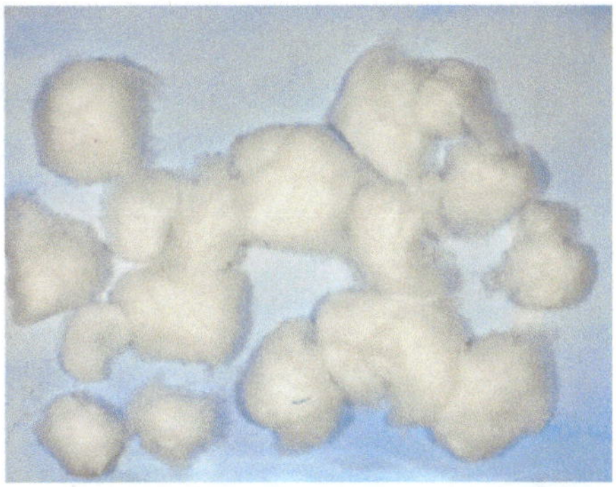
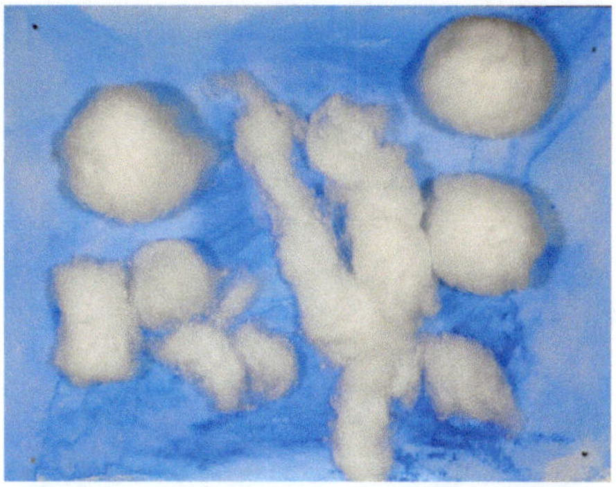
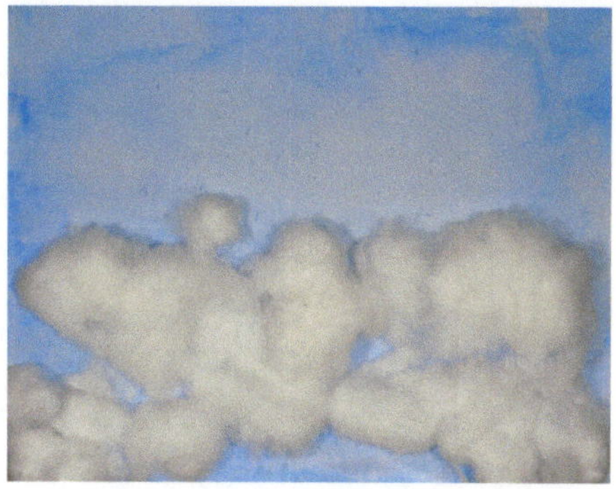
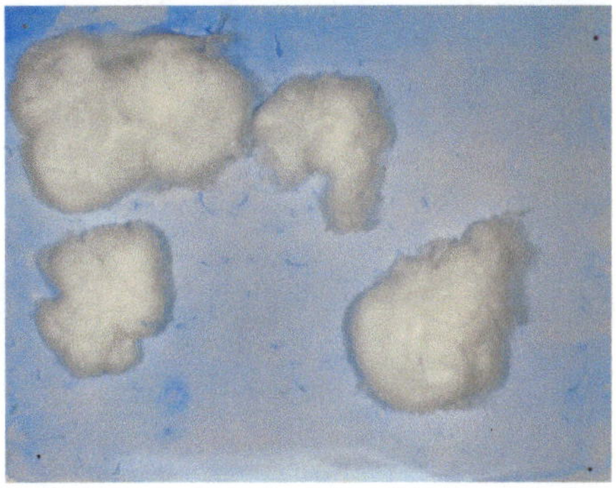
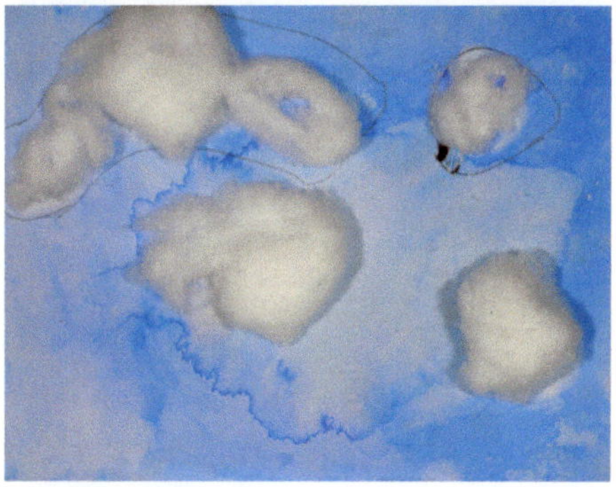
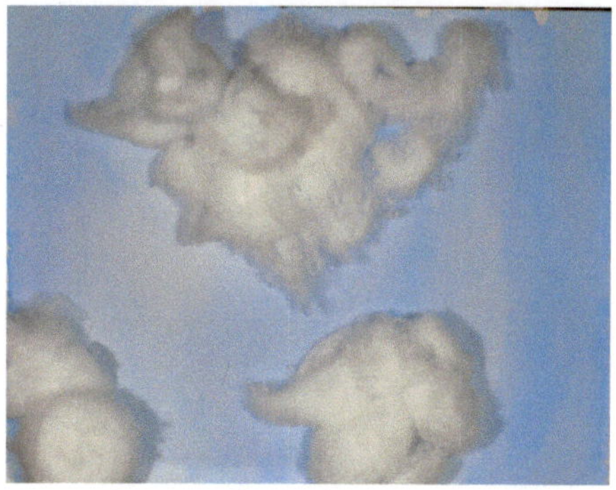
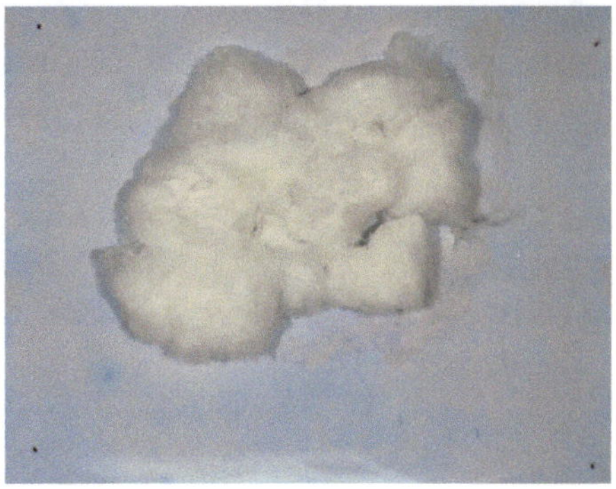
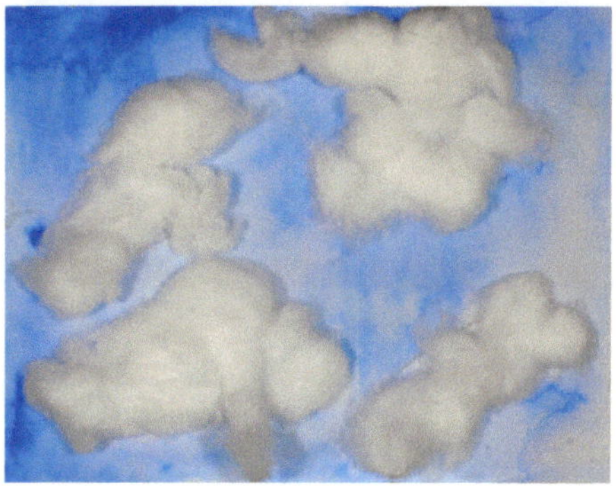
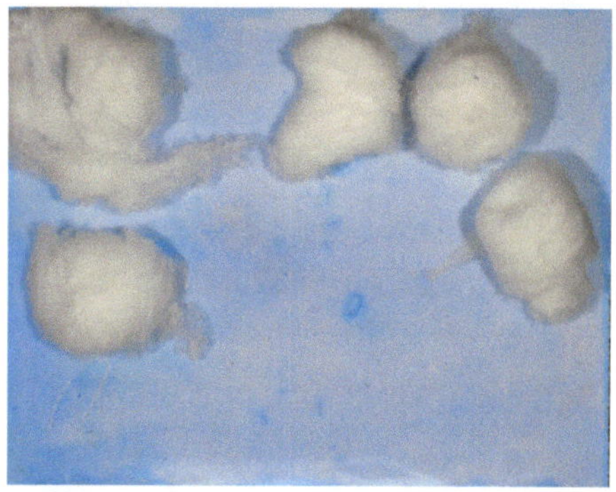
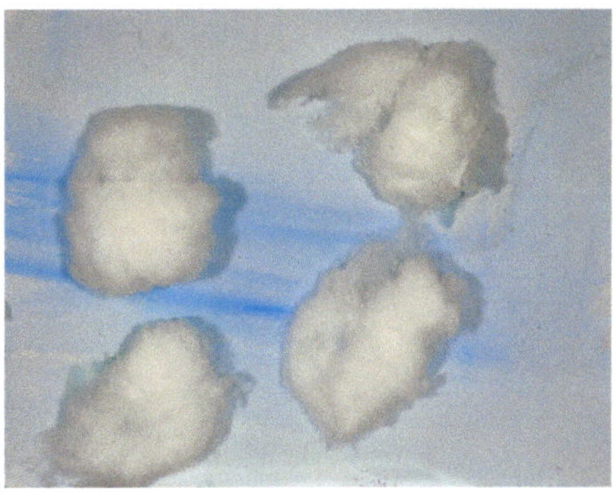
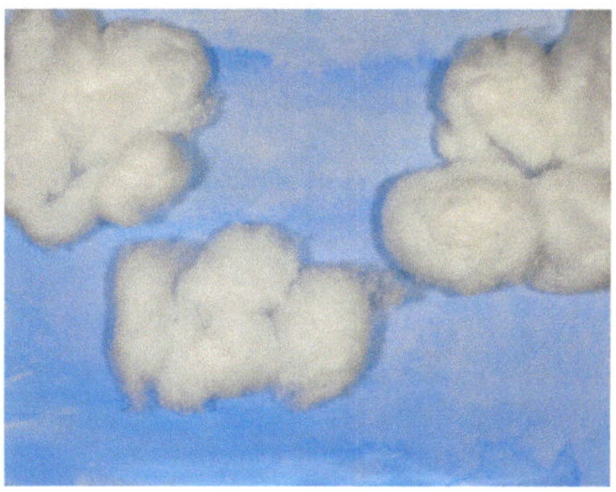
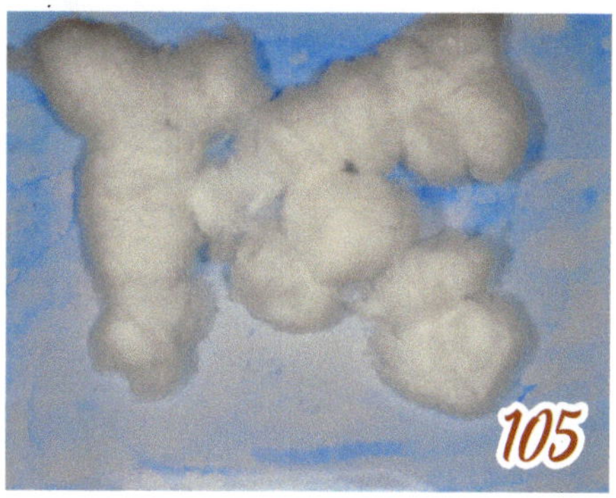

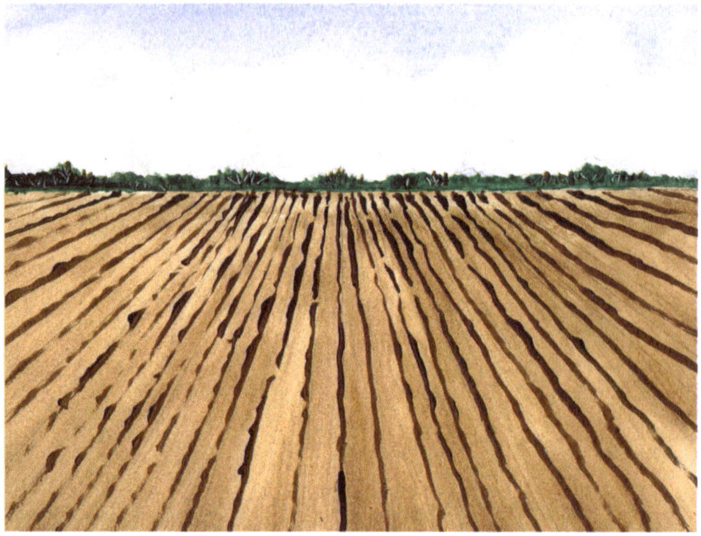

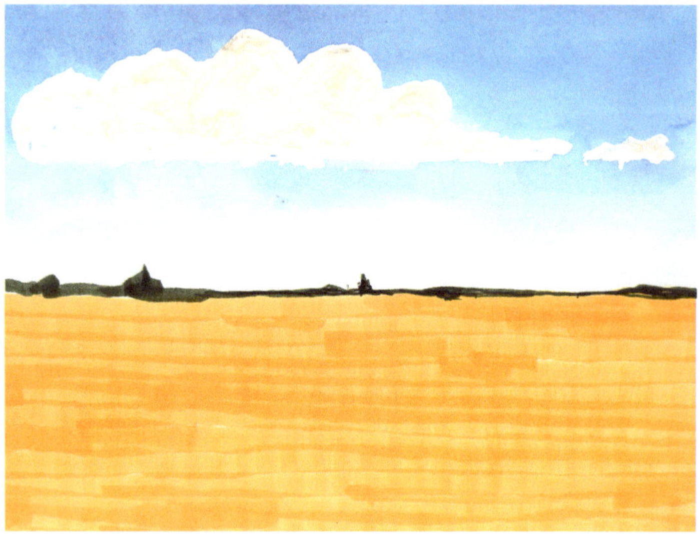

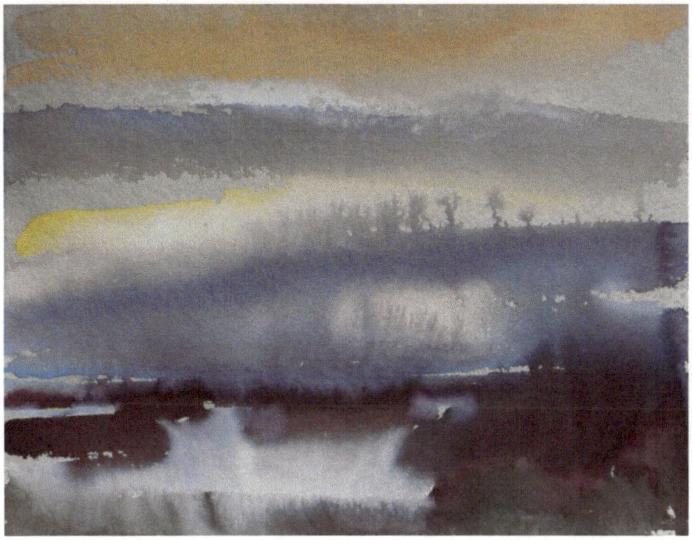

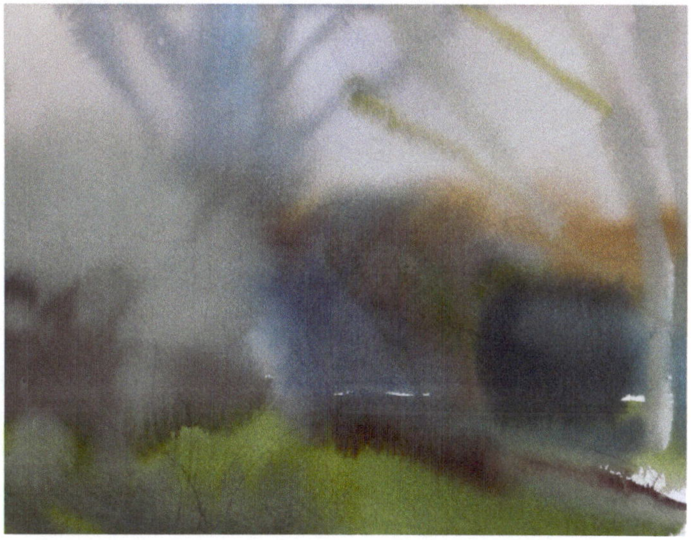

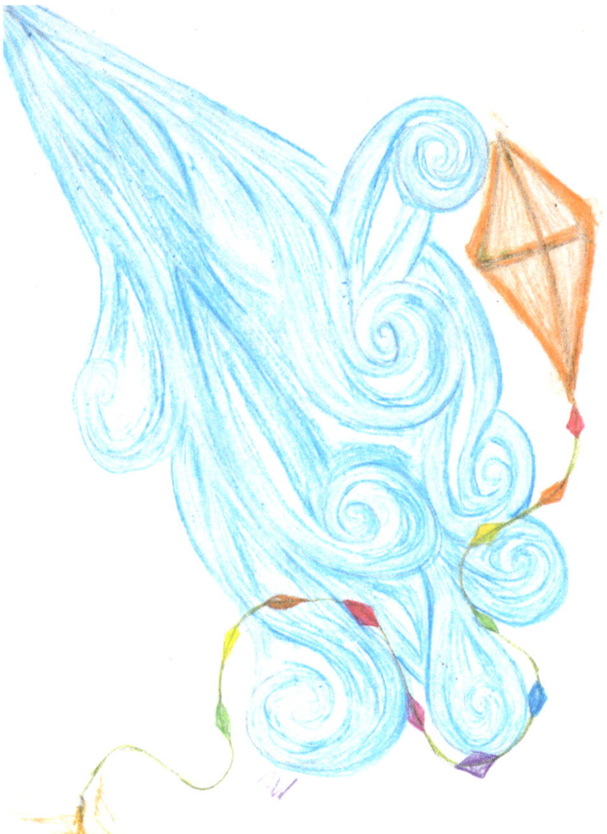

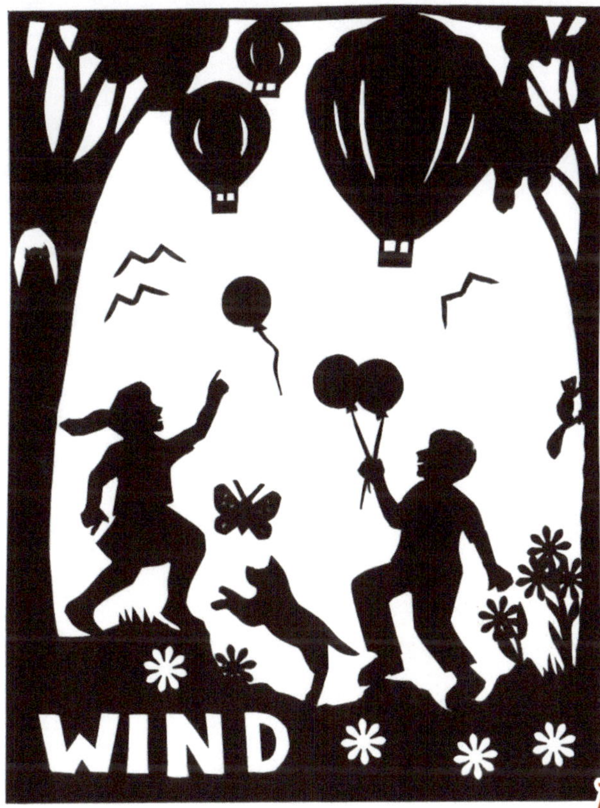
WIND

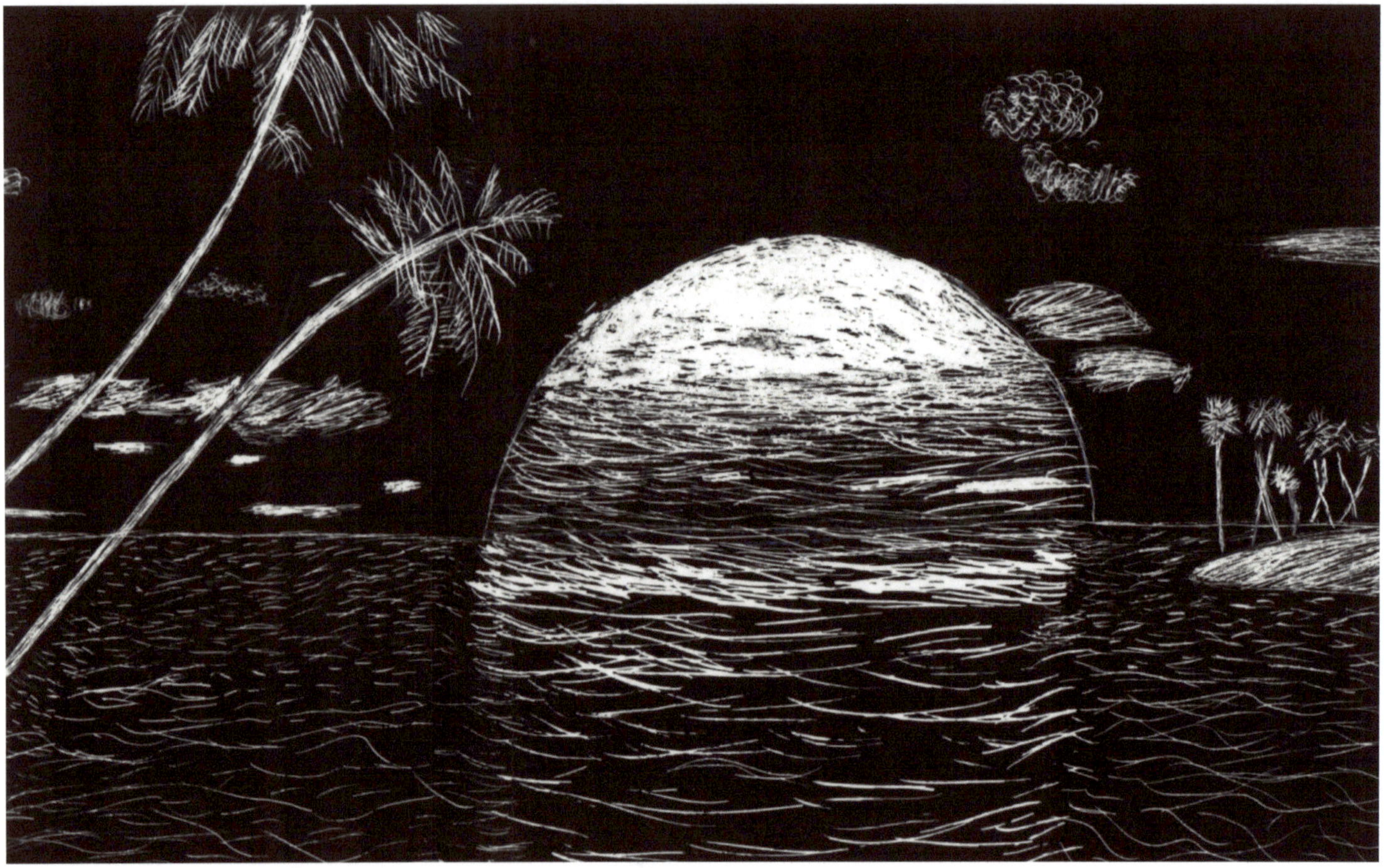

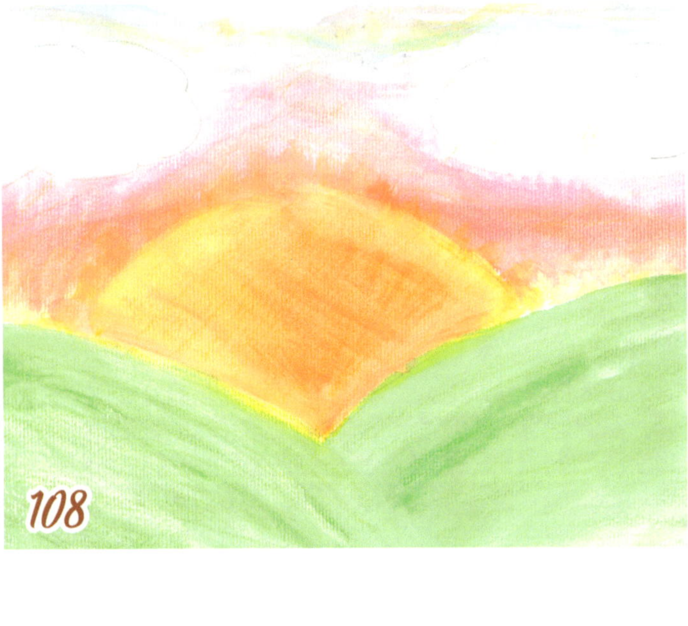

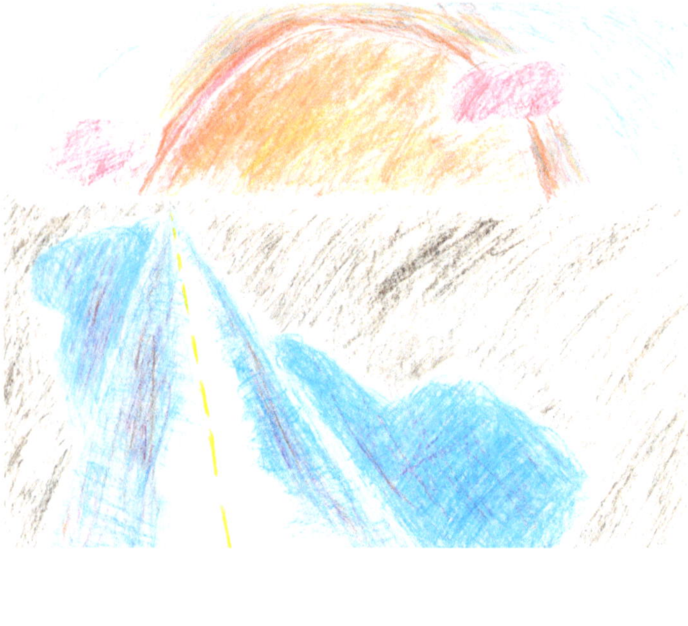

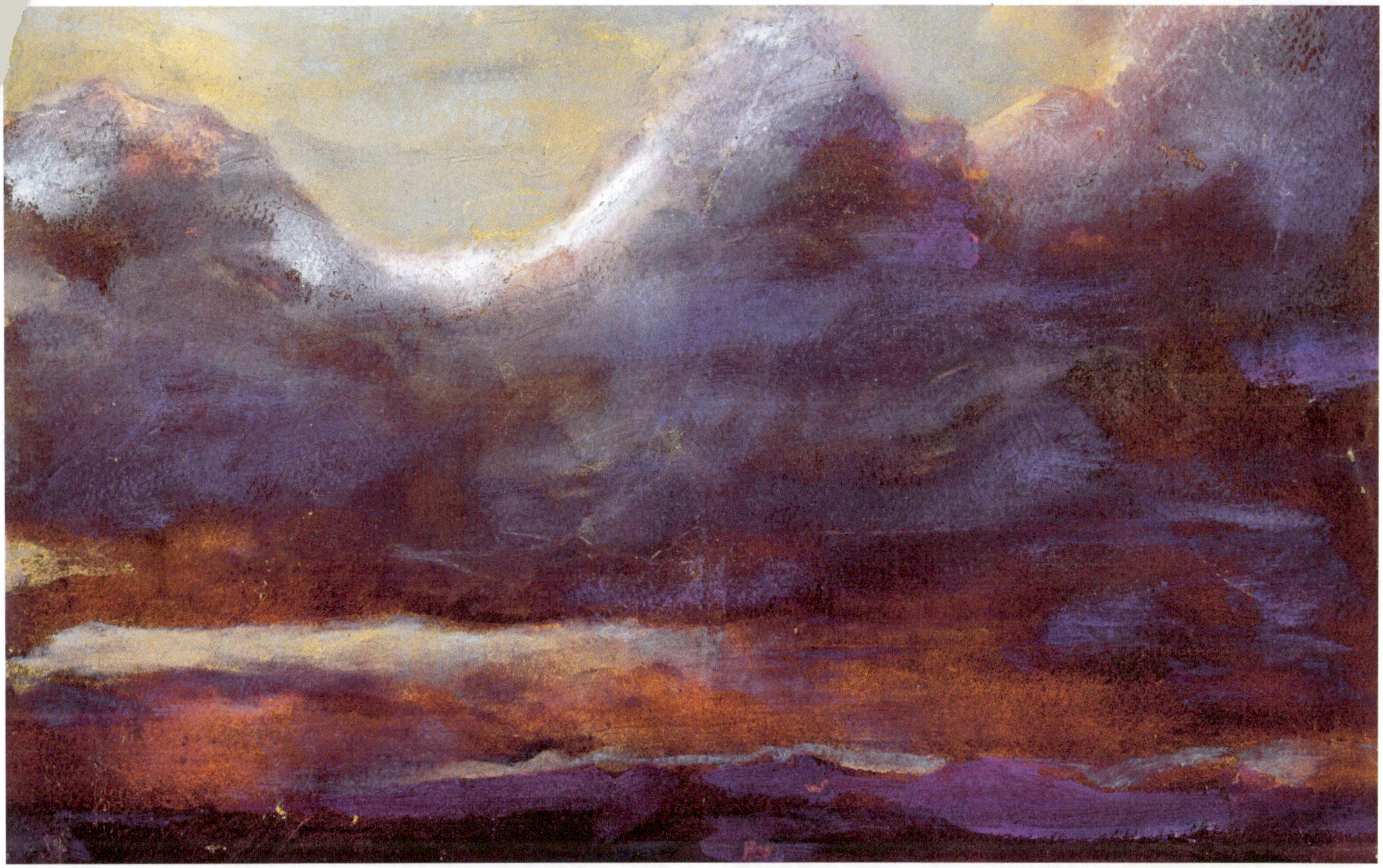

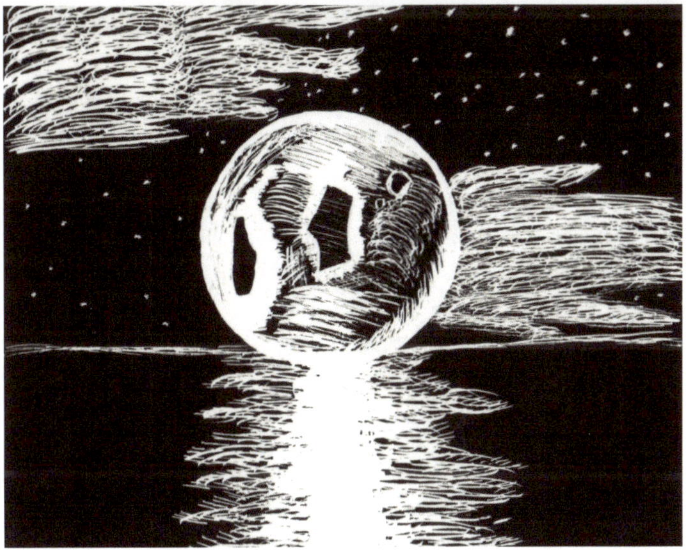

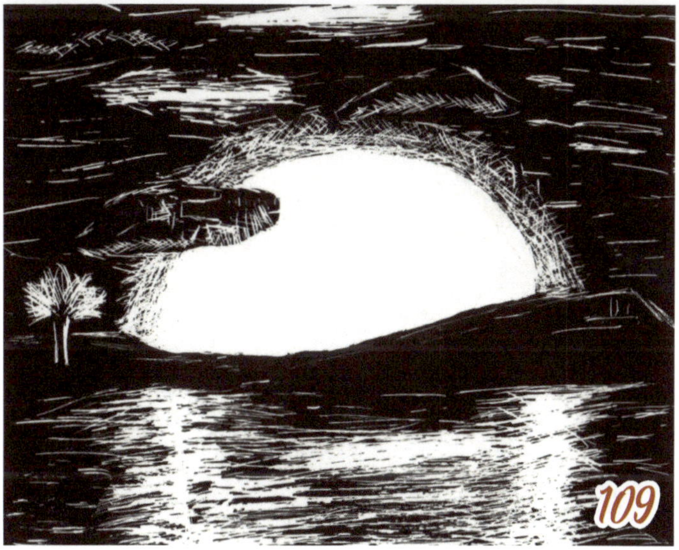

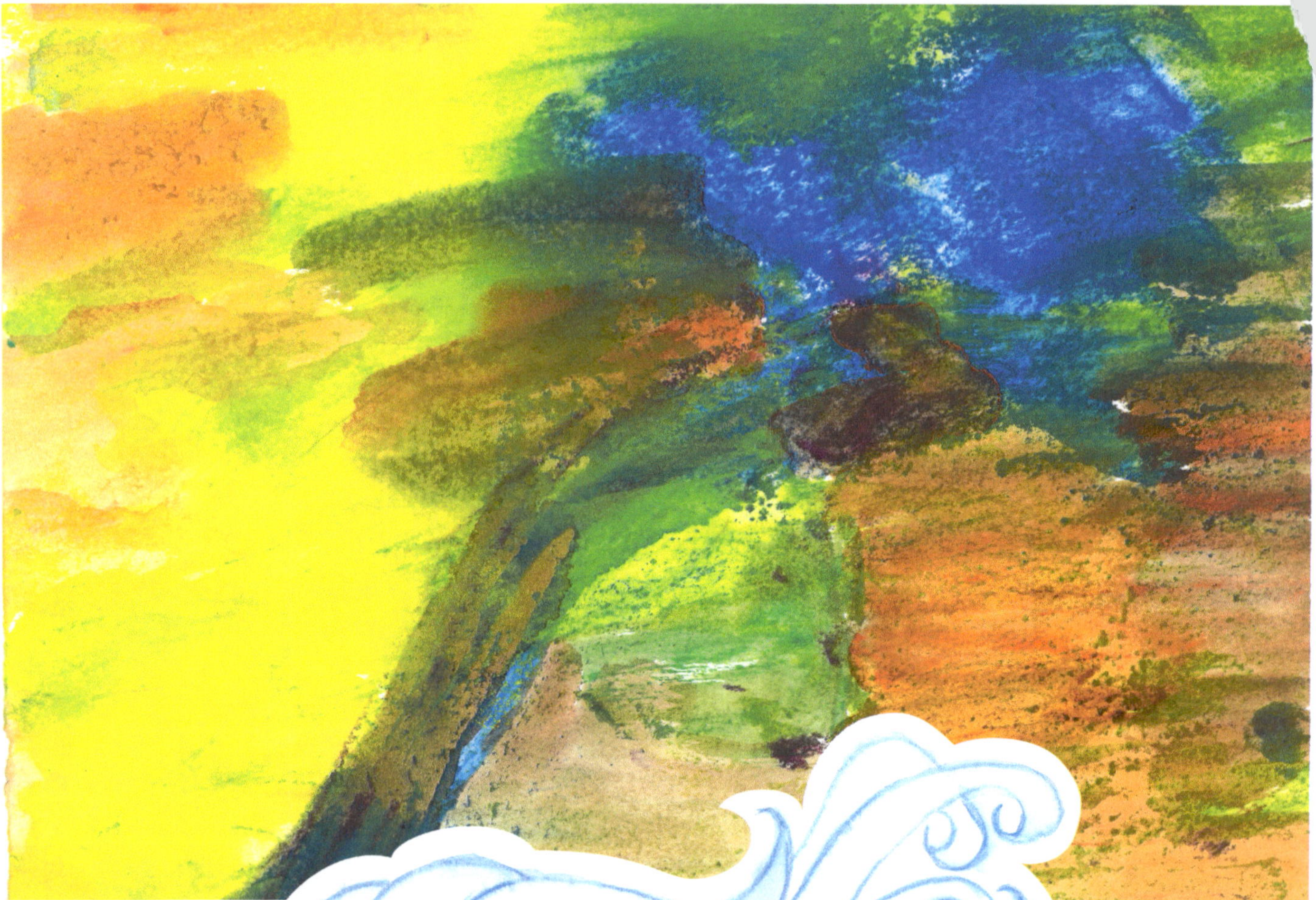

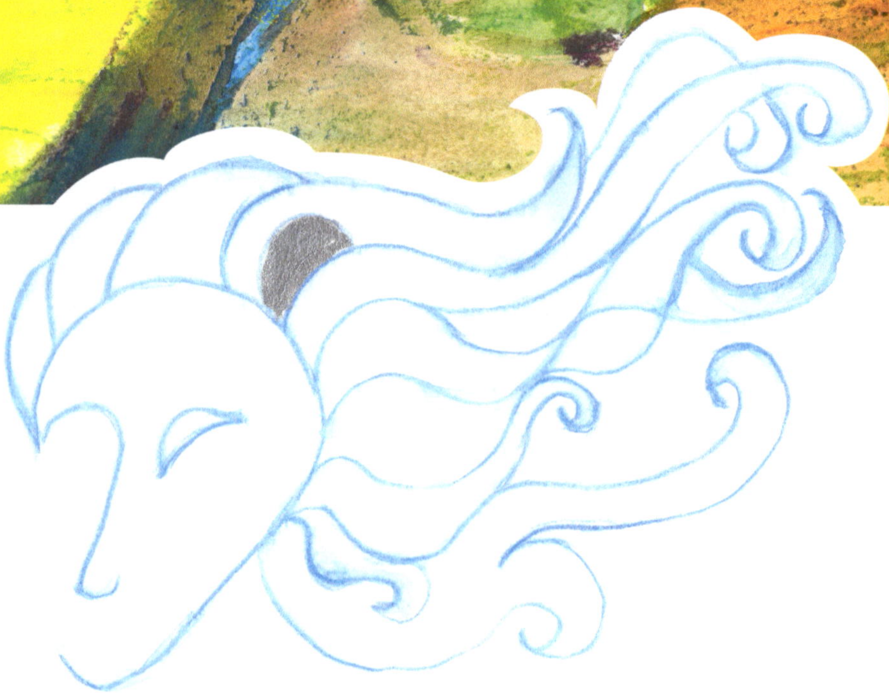

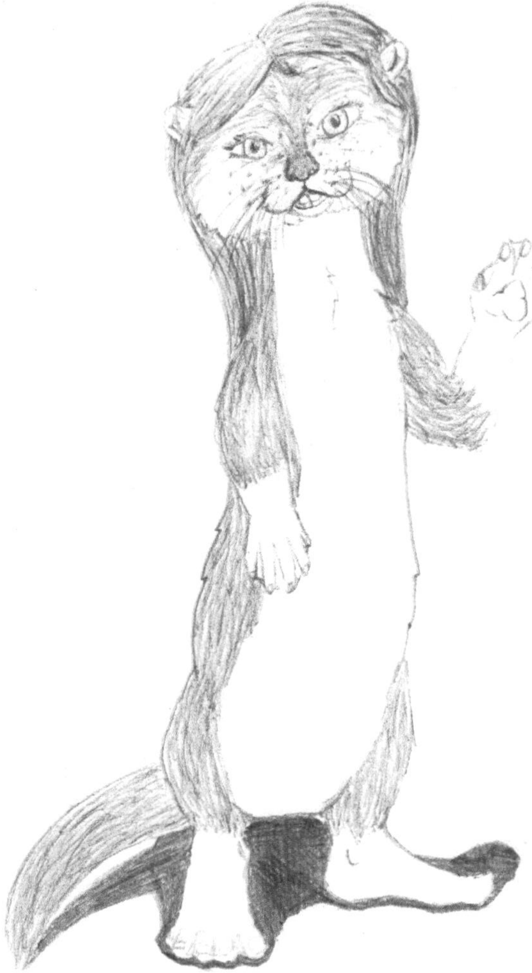

Thank you for your support.

See you around town!

www.ingramcontent.com/pod-product-compliance
Lightning Source LLC
LaVergne TN
LVHW070954210926
848482LV00002B/212
* 9 7 8 1 7 3 6 1 6 9 1 1 7 *